Everyday
Trauma

Everyday Trauma

. . . .

Remapping the Brain's Response
to Stress, Anxiety, and
Painful Memories for a Better Life

Tracey Shors, Ph.D.

FLATIRON
BOOKS
NEW YORK

EVERYDAY TRAUMA. Copyright © 2021 by Dr. Tracey Jo Shors. All rights reserved. Printed in the United States of America. For information, address Flatiron Books, 120 Broadway, New York, NY 10271.

www.flatironbooks.com

Library of Congress Cataloging-in-Publication Data
Names: Shors, Tracey, author.
Title: Everyday trauma : remapping the brain's response to stress, anxiety,
 and painful memories for a better life / Tracey Shors, Ph.D.
Description: First edition. | New York, NY : Flatiron Books, [2021] |
 Includes bibliographical references.
Identifiers: LCCN 2021026489 | ISBN 9781250247001 (hardcover) | ISBN
9781250247025 (ebook)
Subjects: LCSH: Psychic trauma. | Psychic trauma—Treatment. |
 Post-traumatic stress disorder. | Memory. | Brain mapping.
Classification: LCC RC552.T7 S56 2021 | DDC 616.85/21—dc23
LC record available at https://lccn.loc.gov/2021026489

Our books may be purchased in bulk for promotional, educational, or business use. Please contact your local bookseller or the Macmillan Corporate and Premium Sales Department at 1-800-221-7945, extension 5442, or by email at MacmillanSpecialMarkets@macmillan.com.

First Edition: 2021

10 9 8 7 6 5 4 3 2 1

For my glorious son and low-key hero, Evan

Apparently there is nothing that cannot happen today.

—Mark Twain

CONTENTS

Everyday
Trauma

Prologue

Tell me about your despair, yours, and I will tell you mine.

—Mary Oliver, from her poem "Wild Geese"

EVERYONE HAS A STORY

My mother used to say, "Everyone has a story." People tell stories about their love lives, their children, and their professional accomplishments; they talk about challenging childhoods, broken hearts, sexual violence, and emotional abuse. We tell stories for all kinds of reasons: to let others know what has happened to us, and to help people better understand who we are and what our lives mean. But mostly we tell stories to ourselves, stories about what has happened to us in the past, how we feel about what happened, and how we feel about ourselves given all that has happened. We tell stories so that we can learn from our experiences and not repeat those that cause us or others harm. But some of us tell the same stories over and over again, long after they are interesting or of value.

Our brains are designed to create stories from our everyday experiences, and often the stories we repeat are of our everyday traumas. I want to share with you some ways that we can train our brains to tell better stories with the memories we already have and how we might go about creating better memories in the future. But first, I'll need to share with you what I've learned about how the brain generates stories in the first place as we live through the most significant moments in our lives—and how these stories can wreak havoc on our everyday lives by creating feelings of depression and anxiety, panic and fear, and, under rare conditions, a psychotic break with reality. I'll share with you a few stories from my own life and what I've learned about my own brain from personal experience. But this book is not about me. It is about your life and how your brain makes memories of its most poignant moments, good and bad, similar yet different. It is about how these memories influence our present and future. It is about thoughts and how your brain generates thoughts around those memories, day in and day out. It is about the way your brain is ready—always ready—to learn and remember.

Along the way, I hope you will come to realize, as I have, how important it is to keep our brains both mentally and physically fit. We must train our brains not only for what has happened in the past but also for what is happening now and what will surely happen in the future. After all, traumas do occur. People we love will die. We won't always get the friendships that we want and need. We will live through unexpected events like the loss of a job or a car accident or a global pandemic. Things will not always turn out the way we had hoped. My mother was right. Each of us has a story—but with a bit of effort and insight, we can teach our brains not to get lost in or overcome by the memories. We can train our brains to play a leading role in the stories they create.

A UNIQUE STORY OF TRAUMA

Not long ago, I was walking downtown in New York City when I saw a huge mural. On it were painted the words EVERYONE'S DIFFERENT

AND EVERYONE'S THE SAME. It reminded me of a small study in which there were only two participants, a husband and wife.[1] Both had been involved in a terrifying car accident with more than one hundred other cars and even more people. After their car crashed into the massive pileup, they became trapped and were forced to watch a child burn to death in an adjoining car. As they watched in horror, they feared they would die as well. They survived, but as you can imagine, they were left severely traumatized. Four weeks after the accident, the couple was interviewed by a group of psychologists. During the process, they completed a structured interview used by clinicians to evaluate mental health symptoms according to diagnostic criteria. Both participants were diagnosed with acute stress disorder, which is used to describe a constellation of symptoms occurring shortly after a trauma.

The husband reported he had lots of energy during the accident, even managing to break the windshield to make room for their escape, but weeks later, he had trouble concentrating at work, felt on edge, and was irritable. Mostly, he tried not to think about what had happened. And he avoided reminders, especially the highway where the accident had happened. His wife, on the other hand, was "in shock" and "completely frozen" during the accident. She literally could not move and, afterward, said she felt "numb." Later, like her husband, she avoided the highway where it happened, but then she went further and quit driving altogether. She also had trouble concentrating, but compared to her husband, her problems were more severe, leading her to sell her business within months of the accident. She experienced common symptoms associated with trauma—unwanted memories, intrusive thoughts, and distress. But she also reported dissociative experiences, meaning she felt her consciousness "break with reality." Her husband reported fewer symptoms and minimal, if any, dissociation. Prior to the crash, neither of them had mental health concerns, although the wife did have a history of postpartum depression and recounted a traumatic childhood.

Here we have two people, closely aligned both before and after a shared traumatic event but responding in different ways during the event and again after it was over. Now, let's turn to the actual study. The husband and wife were asked to relive the trauma while listening to a script of the accident. During this laboratory experience, the man became anxious, with his heart beating faster—about thirteen beats per minute faster than his normal resting state. His brain was also more active, with increased blood flow to numerous regions, including the temporal cortex, under which the hippocampus and amygdala are located. Both of these brain regions are associated with the creation of traumatic memories. In contrast, the wife said she felt numb while forced to relive the traumatic event, and interestingly her heart rate did not change. Even her brain did not respond as much as her husband's. However, the part of the brain that processes visual information, the visual cortex, was quite active, as if she were perhaps "seeing" the event again with her brain. After the study was over, the couple completed a series of cognitive therapy sessions. The husband recovered within six months, whereas the wife did not fully recover and continued experiencing symptoms consistent with a diagnosis of post-traumatic stress disorder, more commonly referred to as *PTSD*.

This story, albeit tragic, is highly unusual. It is not often that neuroscientists have the opportunity to study the brains of two people who experience the same traumatic event. The couple presumably had similar everyday lives leading up to the trauma—after all, they are married. To be sure, they had different histories and experiences prior to the car crash, but in the big scheme of things, their everyday lives and their lives that day were likely similar. They probably had lunch together and talked about what had happened at work and their plans for the next weekend. Then, all of a sudden, the crash happened, and the man became hyperactive and energized enough to break a windshield (no small feat), while his wife sat frozen in fear. Now it might be tempting to assume that their responses were different because of their genders, but we must be careful about these preconceptions. If the wife had broken the windshield and the husband had sat there frozen, we probably would not explain it away by gender alone. The truth is that we don't know all the circumstances surrounding the accident. Maybe the husband wasn't as trapped as the wife was. Maybe she couldn't see a way out. And we must consider alternative outcomes. For example, what if their daughter or son were in the car? Perhaps they would have responded differently. Parents are known to overcome many obstacles in life in order to protect their children.

The truth is, we can never really know all the differences among individuals that make them respond in one way or another to trauma. We each have a different brain because we each come into the world with different genes and are exposed to different amounts and types of hormones while still in our mothers' wombs. Once we emerge into the world, we experience differing degrees of stress and trauma as we transition from childhood into puberty and adulthood. Each of these life stages comes with its own unique experiences and opportunities for learning that change not only our brains but also the way we use our brains to interact with other people and the world going forward. The way our brains are shaped in turn shapes the way we process our experiences—including our everyday traumas. We, as a human species, are always changing. We are always learning. We always have the chance to reshape our stories for the better.

PART I

The Stories of Our Lives

Life's Traumas—Both Large and Small

> My life has been filled with trauma. When there are a lot of
> traumas in your life, you kind of take them for granted. As far
> as I'm concerned, traumas are everyday occurrences. A looped
> tape of my life runs through my head. I can't seem to forget the
> bad experiences. And there have been so many of them that I'm
> always anxious and worried about what might happen next.
>
> —Kim, forty-one-year-old shop owner

I was recently speaking with a group of people about my work on
trauma and post-traumatic stress disorder (PTSD) when a man in the
group asked me if I was doing any research at the local veterans' hospital. When I explained that my work focuses primarily on women
in the community, he was genuinely surprised. "I thought it was only
veterans who get PTSD," he said. "How does the average woman get
PTSD if she is not in the military?"

His reaction is not unusual. Many of us tend to associate trauma
and PTSD with military personnel. We have read so much about these
psychically wounded warriors that it's easy to forget how many men
and women without military experience struggle with trauma. Living
among us are countless millions struggling to recover from events that
occurred while they were going about the simple business of everyday

life. I'm talking about the friendly next-door neighbor who regularly waves to us as she walks her dog, as well as our mothers, sisters, brothers, daughters, sons, and friends. Why do so many of us feel like wounded warriors, exhausted and scarred by painful past events in our lives? Why can't we stop thinking about what happened? And what can we do to help ourselves recover from the traumatic experiences of life?

Let's start by acknowledging the degree to which trauma is common to all of us. A recent study of nearly seventy thousand people from all around the world reports that more than 70 percent have been exposed to one or more traumas at some time in their lives.[1] Moreover, those who had a history of trauma were more likely to experience additional traumas in their future. For example, persons who experienced physical violence as children were more likely to experience violence as adults, such as muggings or domestic abuse.[2] And as a result, they were at greater risk for being diagnosed with post-traumatic stress disorder. In general, the scientific literature suggests that men are slightly more likely to have traumatic experiences—but far more women suffer from PTSD. In fact, women are two to three times more likely than men to be diagnosed with PTSD.[3] This statistic is alarming and one of the reasons I became interested in how women specifically respond to stress and trauma. Moreover, traumatic experiences don't just produce symptoms of PTSD; they often produce symptoms of depression and anxiety, as well as high blood pressure, insomnia, and even obesity. They also change the way we think and what we think about. I will discuss in depth all of these responses to trauma, but before doing so, we need to consider what we mean by *trauma*. What is it, exactly? And how is it different from stress?

STRESS VERSUS TRAUMA

I have been studying *stress* and *trauma* for several decades, and yet still find it difficult to define them and to distinguish them from each other.[4] The average dictionary definitions of *stress* mention "pressure" or "tension" that is applied or exerted on an object. *Merriam-Webster's* defines stress as "a force exerted when one body or body part presses on, pulls on, pushes against, or tends to compress or twist another body or body part." In this case, the word *body* does not necessarily refer to the human body, but it can be useful to think of it in this way. We might think of psychological stress as a force exerting itself on our brains—twisting them, causing a new, different, and generally unwelcome set of responses. We know it when we feel it.

The word *trauma* is derived from the Greek word for *wound*. If you have been psychologically traumatized, you have been in this sense "wounded." Think about your own life. Have you felt traumatized by any of the events during it? Has anything happened that caused you to feel "unsafe"? Do you have memories of experiences that left you feeling significantly less than whole? Traumatic events are typically described as experiences during which you personally and realistically felt threatened with death or serious injury or harm. But trauma can also arise when we hear about a trauma that happens to someone else, such as the unexpected death of a loved one, family member, or close friend. Traumatic events are wide-ranging and all-encompassing, from car accidents and physical abuse to violence on the streets, natural disasters, and medical conditions. It's hard not to recognize that daily life itself has the potential to be deeply traumatizing.

There are three key differences between stress and trauma: (1) length, (2) intensity, and (3) how we feel at the time and afterward. Let's start with length.

Stress is often categorized as short, as in *acute stress*, or long, as in *chronic stress*. An acute stressor would be something like a small fender bender, a really bad date, or an unfortunate meal you remember best for the stomachache that followed. In contrast, chronic stress goes on and on and on, day in and day out. Chronic stress can be caused by a wide range of circumstances—everything from an overly demanding job to trying to manage unpleasant living conditions or enduring extremely difficult relationships with family or romantic partners. Long-term illnesses, such as cancer, HIV, and long-haul COVID-19, are chronic stressors. Systemic racism also falls into this category, as well as discrimination on the basis of sexual orientation, gender, and age. As one of my students once told me, "In some way or another, every day is sort of a stressful day for me."

In contrast, traumas are usually short in duration, or at least remembered that way. We tend to remember a "car crash," "assault," "earthquake," or even "romantic betrayal" as one event—as an episode. Some clinicians further distinguish between simple and complex trauma. In this case, a *simple trauma* refers to a single event that is definable, such as a car accident. I prefer the term *acute* over the term *simple,* because there is nothing really simple about trauma, regardless of what happened. *Complex trauma,* as the name implies, has more to it—more episodes, more responses, more experiences—and is oftentimes interpersonal in nature, such as the trauma associated with child abuse.

Now for intensity. Traumas are generally more intense than stressors and therefore cause more harm and injury. Recall that the word *trauma* arises from the word *wound,* whereas the word *stress* is defined as a twisting—a remarkably accurate description for the way we feel when we are under stress. Stress twists and turns us around, making us uncomfortable and generally unhappy. But with trauma, there is a wound, and that wound, if serious enough, will be resistant to healing and may even cause irreparable damage.

Finally, for our feelings. In general, traumas elicit negative feelings, whereas stressors do not necessarily. Many stressful events in our lives feel good—graduating from high school, getting married, or landing a dream job. I heard a lecture once during which the teacher described a set of responses and feelings that he was having—heart racing, sweating, anxious thoughts, and so on. He asked us to guess what he was experiencing. We all guessed bad things. But we were wrong; he was describing how he felt when he first fell in love. When we get excited in this way, our body physically responds with similar sensations that we have when we are facing more negative experiences. Apparently the astronaut Neil Armstrong landed on the moon with his heart racing at 150 beats per minute—which was about 80 percent of his maximum. He was stressed but in an excited, good kind of way, albeit with some fear mixed in. So, we can be highly stressed without feeling traumatized. But we cannot be traumatized without also being stressed. Highly traumatic experiences take our stress responses to another level.

EVERYONE IS THE SAME AND DIFFERENT

Each person responds to significant life events a bit differently, and these so-called individual differences are important to keep in mind when trying to understand any one person's response. Recall the car accident I described earlier, during which the husband broke the windshield while his wife sat frozen in fear. Why did the two of them respond so differently? Or take 9/11. When the Twin Towers collapsed, some who witnessed the event ended up with PTSD, while others did not. Why were some so traumatized that they moved out of New York City altogether, while others felt fine going back downtown to work? Also, one person's memories may be different from or sharper than someone else's. I have a girlfriend who seems to remember everything—even waiters we had in a restaurant years ago, whereas I have no recollection whatsoever. And people are similarly different when it comes to memories of trauma. While someone can remember what happened vividly, another person may shut the memory out entirely.

Individual differences are often attributed to differences in resilience. It helps to think of each person as a rope. One person might be born with a strong, thick rope, while another is born with a thin, more fragile one. Someone who ended up with PTSD after 9/11 might have been born with a more fragile rope or lived through circumstances that frayed the rope, making it more vulnerable even before that day. If you have a stronger rope or can cultivate one, it will take more stress to break it. But if you have a fragile one or if life has weakened your rope along the way, a smaller amount of stress can cause it to fray and maybe even break. We can make our ropes stronger with healthy behaviors, such as physical exercise, eating well, and sleeping well. But we should also go beyond the obvious. We should and can train our brains to learn new skills that will help us accept and live with our own thoughts and memories—more on how to do this later. Then, when stress and trauma arise, our ropes will be less likely to break.

Take Antonio, who was a standout athlete his whole life, even though he suffered with dysthymia, a mild but persistent form of depression. When he broke his leg, his life came to a dead stop.

The whole time that I was locked up in my room I was alone with my own thoughts. I had already suffered from some depression, but it started to become more noticeable. I hit the lowest point in my life, and at that point, I truly felt my life came to an end, and I didn't care what happened. I had pushed away my friends, I wasn't eating, and I was sleeping all the time. I was very moody and unhappy and did some things I do not wish to mention. Do I blame my injury for my depression? No, but I do believe that the injury triggered tough emotions. As much as I wish the injury never happened, I believe that if it hadn't, my depression would have been less noticeable and who knows what the outcome could have been. Because of my trauma, I am alive today.

It is needless to say again, but I will: we are all different. What is stressful or traumatic to one person may not be so to another. Life brings to each of us differing experiences and, as a result, differing responses. But we are also very much the same. We are human beings living in a life composed of constant change and potential danger, interweaved with opportunities for new learning and the certainty of still more change.

COMMON SOURCES OF STRESS AND TRAUMA

In the course of living an average life, we are all going to experience traumatic events, some of them severe enough to cause PTSD and others not. It's estimated, for example, that regardless of military service, one out of ten women will develop PTSD at some point in their lives.[5] The risk is less for men but they are still vulnerable. So, now I am going to review some of the common stressors and traumas that people experience. This list is not meant to be exhaustive; that would be impossible. But I do want you to at least recognize some of the traumatic life events that you or someone you know might have experienced but maybe would not have considered traumatic at the time or in retrospect. All that said, I do not mean to imply that all perturbations in life are traumatic. Years ago an established figure in the trauma field told me he was concerned about the overuse of the word *trauma* and, in particular, the increase in diagnoses of trauma-related mental illnesses such as PTSD. I remember him telling me that someone came to him with symptoms that they said they had acquired while watching the nightly news on television. But on the other side of the spectrum are those of our fellow human beings who experience so many traumas that they can barely recall them, much less categorize them. So with those caveats in mind, let's review some common stressors and sources of trauma.

Difficult Childhoods

Not everyone has a perfect childhood. Many people grow up in homes that don't feel safe, with caretakers who are physically and/or emotionally terrifying. Ongoing verbal abuse from a partner or a parent takes a toll on our mental and emotional well-being and in this way can be deeply traumatic. When these experiences continue day after day, children often learn to expect still more misfortune. As adults, these individuals may approach the world as if they were expecting to be hit or verbally assaulted—ready to flinch, run for cover, or completely shut down. A difficult childhood can also include growing up with food insecurity or in neighborhoods that feel dangerous and scary or with the experience of being repeatedly bullied. Others remember parents who abused substances, such as drugs or alcohol, and whose personalities changed accordingly. A large number of people talk about being afraid of siblings who were abusive, while others were traumatized when their parents were divorced. And some childhood experiences are almost too tragic to comprehend. One of my close friends was traumatized throughout his childhood by his own father, who suffered from severe mental illness. His father once tried to run him over with his car![6]

Accidents

Most people are involved in an accident of some sort in their lifetimes, and although they may not be as dramatic as the car accident described in the prologue, they nonetheless can leave us anxious and afraid, with racing hearts and sweaty palms. And there are other possible repercussions, such as self-blame and guilt even if the person is not at fault. As people age, they are more likely to fall and injure themselves, oftentimes breaking bones. These types of accidents can spiral downward into a loss of overall health and well-being.

Disease and Illness

Being diagnosed with a serious medical condition is often traumatizing. A person may spend a year or more undergoing a variety of medical procedures and receiving treatments ranging from surgery to chemotherapy. Imagine lying in bed every night thinking about what could happen. Such experiences are highly stressful and often traumatic. Many people don't have access to health care or medical insurance and must make impossible choices about their care or the care of their loved ones. There is still so much we must learn about the long-term consequences of suffering from a serious illness even if the illness is successfully treated, as we have learned about COVID-19 during the pandemic.

Life-Threatening Events

Natural disasters such as floods, hurricanes, and tornadoes make us more vulnerable to PTSD. Someone I knew was nearly hit by a television that flew off the counter during one of the larger earthquakes in Los Angeles. Glass was everywhere, and although he was not physically injured, he said he didn't feel safe living in an earthquake zone anymore. He moved to the East Coast, only to be caught up in the destruction of Hurricane Sandy.

First Responders, Emergency Medical Personnel, Relief Workers, and Other Health Care Workers

Police, firefighters, nurses, doctors, and other medical personnel are regularly exposed to trauma—sometimes so much so that they become traumatized themselves. It's hard to overstate the trauma experienced by first responders called to help people who are literally gasping for air after being infected with the novel coronavirus. Reporters on the beat are also vulnerable.

Extended Periods of Caretaking

It's difficult to appreciate the degree of stress that must come with caretaking, not to mention all the worry and concern—and the guilt. "Am I doing the right thing? Should I be doing more?" It is difficult when the care is appreciated and perhaps more so when it is not. The number of people who reach old age is increasing every year, and with that comes more and greater need for 24/7 care, either at home or in long-term care facilities. We need only remember all the people who were left alone in hospitals and nursing homes during the pandemic, isolated from those they loved and needed the most.

Death of a Loved One

People often say that losing a loved one is just part of life, but that doesn't make it any easier when it happens. Losing your parent is traumatic enough, but how does anyone ever recover from the loss of a child? And what about the trauma of suicide, with so many questions left unanswered?

Romantic Love and Breakups

We all know people who have been traumatized by romantic breakups. I met a woman the other day who, within minutes, told me all about her recent breakup. Out of the blue, her boyfriend of eight years left her. She couldn't eat or sleep and spent most of her days crying. She was moving across the country just to get away from the memories. Another woman told me about how she had just met the man of her dreams. They were walking out of a restaurant one night and someone came up and stabbed him to death. She then had to call his parents to let them know what had happened to their son. She was so devastated that she quit her job as a lawyer.

Violence

Think about the children and teachers present at any of the school shootings that have taken place in this country. Or those who live in neighborhoods where gun violence occurs on their street corners. Others experience sexual and physical violence, with as many as one in three women worldwide being victimized in their lifetimes.[7] And the trauma of domestic violence, much of it committed in secret with no end in sight, cannot be overestimated.

Pregnancy and Childbirth

Some women are traumatized by accidental pregnancies, and others while trying to become pregnant, which is not always an easy process. And how about childbirth itself? Women tend to expect childbirth to be an idyllic experience filled with nothing but joy, but even today, it doesn't always work out that way. Women still experience difficult labor, extreme pain, sometimes combined with fears for the baby's well-being. Many women describe difficult circumstances— epidurals that don't work, neglectful hospital personnel, and partners who don't show up.

Witnessing a Loved One in Danger

Watching someone else experience trauma can be traumatizing. People who have partners or children with ongoing emotional problems or a difficult psychiatric diagnosis often refer to their day-to-day lives as traumatic, as do parents who are worried about their child's addictions. I know a woman who has two adult children in rehab. She says, "Every time the phone rings, I go into panic mode. What next?"

Poverty and Homelessness

Think about the person living day-to-day in poverty or, worse yet, homeless with no food to eat. One meta-analysis (which combines the results from multiple studies) reported that a large number of people who are homeless would meet diagnostic criteria for PTSD.[8] And, sadly, most of them assume their problems can all be explained away by their homelessness. To make matters worse, they often blame themselves for their living conditions.

Gender Discrimination and Racism

The word *trauma* tends to make people think of war or a big natural disaster, such as an earthquake. However, untold numbers suffer with the everyday trauma of discrimination while being denied educational and employment opportunities as well as medical care. Systemic racism, sexism, and homophobia are notably traumatic and take their toll on people all over the world. Many live day in and day out with everything from implicit bias to full-blown violence, simply because of the color of their skin or their sexual orientation.

MY FIRST BRUSH WITH REAL FEAR

In my own life, I've been fortunate to experience few traumas. Yet when I think about personal traumas and the potential impact they can have on behavior, I remember one very scary episode. It was a short event, and the activation of traumatic symptoms did not last for long. Yet the memories of the experience had a lasting impact on my life and continue to influence some of the decisions I make and how I behave to this day.

It was the late 1980s. I was living in California in a house in the Hollywood Hills—right up by the HOLLYWOOD sign we've all seen featured on dozens of TV shows. Built along the side of a cliff, it was a cool house—with glass walls and big windows, several levels, and some exterior stairways. Perched there, alone, with no other homes nearby, it looked like it belonged on a TV set. The closest buildings were in neighboring Griffith Park, a large open area that features an observatory. There were some people walking around during the day, but at night it was pretty much deserted.

On this night, I was in the house alone and had started out the evening in the master bedroom on the top floor, where I was reading, studying, and briefly talking on the phone. The house, which didn't have air-conditioning, normally cooled down right after sunset, but on this particular evening the heat lingered. I got ready for bed but, because it was so hot, decided I would have an easier time sleeping if I was down on the lowest level, where there was a tiny bedroom with one window and a small twin bed.

When I reached this room, it was about midnight; I was too tired even to read, so I quickly got into bed, turned off the light, and headed for sleep. There was little moonlight that night, and it was extraordinarily quiet. Almost as soon as I closed my eyes and started to doze off, I heard a noise coming from an outside staircase. At first, I didn't think much of it. Maybe it was a deer or a coyote. Then I realized how unlikely it was that a deer or a coyote would be climbing a staircase. Suddenly, I was wide awake. There was a loud banging—somebody was trying to break down a door. I was able to look out the window and could see the back of a figure, wearing dark clothing, on the exterior staircase. I couldn't tell where he was going, and I wasn't about to try to find out.

All the clichés come to mind: I was too scared to move, gripped by terror, paralyzed by fear. I remember trying to think about what I *could* do. Could I get to the phone? Could I run for the front door? But any action seemed futile and likely to draw more attention to my location. This was before the time when we all carried cell phones. There was no landline in the room. I didn't think I could get to the car; my purse and car keys were in another room. I thought about trying to get to the closet, but I was frozen in fear and couldn't move. So, I just lay there for hours, my heart pounding so loudly that I was sure this person, whoever he was, could hear it. Straining to hear, I heard nothing. I didn't have any idea where he was or if he was still trying to find a way in. When the first bit of sun started to come up and I could finally see, I managed to get my keys and run to my car. I went to my best friend's house, where I woke her by banging on her door. I don't remember much of what happened after that.

In my relatively sheltered life, I had never been afraid like that before. And from that day forward, I did things differently. Whenever I rent an apartment, I make sure that it is not on the ground floor, and wherever I live, I have either a large dog or a serious security system. I am more vigilant when out at night—avoiding places that are dark or dangerous or unknown, and I'm always looking around to make sure that no one is following me. These types of responses—vigilance and avoidance—are normal and good for us most of the time. Our brains are designed to elicit them. However, it is possible for our brains to get carried away, and before we know it, we are avoiding not only the people, places, and things directly related to the traumatic event but also life in general. Trauma memories are powerful. They form quickly and can take over our thoughts and actions as we progress through our lives. And if they can't take them over entirely, they will surely try.

2

How Stress and Trauma
Change Our Lives

I am a survivor of sexual trauma as a young child. I'm now
forty-five and have been in therapy for years. I think of the
incident as a grain of sand in an oyster. Everything after the
trauma somehow grows around it and is formed by it.

—Email from Olivia

Each year, I teach hundreds of college students about stress and mental health. Some years ago, I gave them this assignment: *Think about
the most stressful event in your life and then write a few paragraphs about
how you responded psychologically at the time and then a few paragraphs
about how your body felt.* One student wrote of driving alone to another state to have an abortion. Another was working as an EMT
when he received an emergency call from his own address, where his
father was having a heart attack. Many wrote about sporting or car
accidents from which they had difficulty recovering. Others watched
parents and grandparents die. Still others felt traumatized by romantic breakups with partners they trusted. But this was not the assignment. In the instructions, I explicitly tell them not to include any
defining details of the event, but apparently they find it easier, or they
prefer to write about *what* happened—when and where and who was

there—and not necessarily about how they felt or what they were thinking at the time.

This is true for most of us: we want to tell the stories of our lives but don't necessarily want to relive the emotions we had at the time or consider how the memories continue to haunt our everyday thoughts in the present. In fact, it is not always healthy to relive those feelings or to think too hard about those memories. But if we want to live more peacefully alongside stressful thoughts and traumatic memories, we need to become more aware of them, as one person described:

> What can I possibly do to feel normal again? That day became a turning point for me. It is as if I were on a roller coaster and suddenly it stopped. I am stuck on the roller coaster, and I am not able to move or get free. I did not expect the roller coaster to stop, and I am not ready for this. All I know is that it stopped, and now I am waiting for help until I get back to safety. This feeling of helplessness and confusion and fear and maybe more is what I am able to relate to. I am at a standstill.

From this level of awareness, we can begin to move forward.

PTSD—WHEN TRAUMA SETTLES IN FOR THE LONG HAUL

Trauma is a complicated term and concept, but clinically, it is most often associated with PTSD, and so this is where I'll begin. A diagnosis of PTSD generally comes after a "structured" interview, which is conducted by a medical professional, such as a psychologist or a psychiatrist. The questions are organized around criteria outlined in the *Diagnostic and Statistical Manual of Mental Disorders,* otherwise known as the *DSM*.[1] This manual has been through numerous editions since its inception in the 1950s, and along the way the criteria for each diagnosis, including PTSD, have changed—but its purpose has not. The *DSM* is heuristic—meaning it is a practical tool that helps clinicians speak the same language and come to a consensus about what they are hearing from clients and from other clinicians with respect to mental illness. It is also used by scientists so that they can compare and contrast findings across laboratories and universities. The *DSM* is not perfect, and it does not do everything. It was created for diagnosis and therefore does not address the underlying cause of a disorder or syndrome. Nor does it suggest or prescribe medications and treatments.

To be diagnosed with PTSD, someone must report symptoms and experiences that meet a set of criteria. First, we consider the actual traumatic event. Someone must be exposed to death, be threatened with death, experience serious injury, or be threatened with injury in one of the following ways:

• Personal exposure to the trauma
• Witnessing a trauma
• Learning that a loved one was exposed to trauma
• Indirect exposure to a trauma (as occurs for first responders and medical personnel)

In response to traumatic events, people experience a number of symptoms, only some of which are necessary for a diagnosis of PTSD. The "intrusion" symptoms include intrusive thoughts, during which people often revisit unsettling memories of what happened and experience emotional distress while doing so. These experiences can come in the form of flashbacks or nightmares and often invoke some kind of behavioral response, such as jumpiness or startle responses. Other symptoms relate to avoidance, during which someone avoids reminders of the traumatic event, such as the people who were there or the places where it happened. Recall the husband in the car accident who avoided the highway where it happened, while his wife did not want to drive at all. Avoidance is a common response and yet can be difficult to identify. When someone starts to avoid reminders of a trauma, they often generalize and start to avoid more than just the actual reminder (such as the highway). They may avoid people in general or stop going out in the world. Some people do not even notice they are doing it or how isolated they have become because of their own behavior.

A diagnosis of PTSD also includes changes in cognition and mood, such as depressive feelings, which may include blame and guilt. A woman named Riley still feels guilty about an experience she had with her grandfather, who died unexpectedly and in a tragic way years ago.

> I remember one specific time that I made him upset and have never seen him so angry. Thinking about that memory makes my stomach go into knots and makes me feel like a horrible person. I was too young of a person to realize how bad of a choice I made at the time. Even though he said he forgave me, it drives me crazy that I cannot go back and change what I did.

After experiencing a trauma, people might lose interest in activities and have trouble concentrating or remembering what happened. They may have problems falling and staying asleep. They often don't feel like doing things that they used to enjoy, and they may find themselves reacting more than they normally would to everyday life, perhaps lashing out in anger or having unprotected sex, gambling, or engaging in alcohol and drug abuse. They may forget key features of the trauma while having negative thoughts about themselves and the world.

Importantly, a diagnosis of PTSD only occurs if the symptoms persist for more than a month and cause functional impairment. In other words, the symptoms must be severe enough that the person cannot go to work or school, take care of the family, or function in their everyday lives. Moreover, the symptoms cannot be attributed to the side effects of medications, substance use, or other medical conditions. As you can see, there are many potential responses to traumatic life experience, some of which lead to PTSD and others that do not. In fact, a large majority of people who experience traumatic life events do not develop PTSD. We did a study some years ago with women who had experienced sexual trauma, mostly as teenagers and young adults. About 30 percent of the participants were diagnosed with PTSD when we interviewed them, which means that nearly 70 percent were not despite the severe trauma they experienced and the many symptoms they felt in the immediate aftermath.[2]

Let's now delve deeper into two common responses to trauma—anxiety and depression—and a less common one—psychosis.

ANXIETY IN RESPONSE TO TRAUMA

Let's start with anxiety. *"My anxiety affects everything I do."* Lisa, a middle-aged woman from Rhode Island, says that her anxiety and panic attacks began when she was only nine years old. As she remembers it,

> I came home from school to find my uncle sitting at the kitchen table with my dad. I got into my dad's lap, and that is when he told me that my mom had gone to sleep and would not wake up. I pretended that I didn't know what had happened, but deep down, I knew. And I was already traumatized by other deaths in the family, and so when this happened, I lost it. I missed many days of school, just lying on the couch curled up in a ball. My stomach ached, and I could not eat or sleep. Then I began to have severe anxiety about death, asking my father over and over if I was going to die and how did he know that I wasn't going to die. I began engaging in obsessive compulsive rituals to keep me from dying,

such as taking my pulse every few minutes. This went on for nearly three years. My father took me to many doctors, but they just shrugged and said they had no idea what was wrong. Even today, many years later, I suffer with panic attacks, but with the help of a therapist and medication, my anxiety is better.

Anxiety is a common response to trauma. It manifests itself in many forms, most of them felt in the body. The most common one is a sense of dread—as if something really bad were about to happen. To prepare for this, the body gets into physiological states that produce hypervigilance. Vigilance is the feeling of being alert—paying attention. Hypervigilance takes normal vigilance to a state of high alert and a feeling of being on edge. While in this state, even the most innocuous of events can elicit a so-called startle response. For example, before the trauma a car horn outside might cause someone to turn their head or look out the window. After the trauma, the sound of the horn causes someone to physically jump into the air. And it can be further exacerbated, as responses to many normal, everyday sounds and events become exaggerated.

Symptoms of anxiety can go beyond the startle response to include dry mouth, perspiration, and gastrointestinal problems ranging from nausea to sudden diarrhea. Some people experience panic attacks, which can be accompanied by physical pain, trembling, and a pounding heart. When someone first has a panic attack, they often think they are having a heart attack. Because the feeling is so dreadful, people who have one panic attack often start to become afraid of having yet another. As a result, much like the husband and wife who were traumatized by their car crash, they might start to avoid people or places or situations where they think it could happen again. This is often referred to as *vicious cycle* behavior, because their behaviors tend to increase anxiety and the likelihood of having yet another attack.

Many people who have anxiety have what is called *generalized anxiety,* meaning it is almost always around and not necessarily focused on or induced by any one person or circumstance in particular. Generalized anxiety is one of the most common psychological problems for people in the world. Other types of anxiety are more constrained but have similar consequences. For example, people who are agoraphobic become afraid of open spaces, while those with social phobia are afraid of doing things in front of other people. These forms of anxiety are triggered by something specific but can still result in a heightened stress response.

Anxiety is generally felt in the body but can also be accompanied by thoughts in the brain, which are usually about the future, and I don't mean thinking about what's for dinner. Rather, anxious thoughts are more like, *When I think about the future, I can't seem to catch my breath,* or *When I think about doing things with people out in the world, I just want to stay home,* or *Just thinking about going into work upsets my stomach and makes me want to throw up.* Even though most of these thoughts and feelings are about the future, they are felt in the present moment with varying degrees of unpleasant feelings in the body. And as we think about the future, we often compare what could happen to what happened in the past. Indeed, when we are anxious about the future, it is usually because of something bad that happened in the past. As we compare the past to the present and consider the future, the feeling of anxiety arises in our body, and the vicious cycle begins again.

Feelings of anxiety are not *always* bad, though.[3] These feelings keep us alert and ready to do whatever we can when danger arises. In his article "What Good Is Feeling Bad?" evolutionary physician Randolph Nesse suggests that feelings of fear and anxiety evolved in humans and other animals so that we will act, even if it just means moving to a new neighborhood, making new friends, or taking our burning hand off the stove.[4] These feelings remind us and, in some cases, implore us to act. We probably wouldn't have survived as a species without some anxiety. But when it starts to take over your life, it is not good for your life—or your brain.

DEPRESSION ALONGSIDE TRAUMA

Depression is also a common response to trauma. Here is an example. Annette lives alone but wishes she didn't. She spent years battling depression and then experienced two traumas in close succession. First, she lost her house in a fire, and then her live-in boyfriend broke up with her. As she recalls:

> I always tended to get sad and a little bit depressed, but it was never anything serious until Steven broke up with me. Two months after we lost our house, out of the blue, Steven turned to me and said, "I can't do this anymore." At first, I didn't have any idea what he was talking about. I still can't believe that he left me, especially given what had just happened.
>
> The fire combined with Steven's "betrayal"—that's the only word I can think of—was paralyzing. I forced myself to go to work every morning and do my job. And then I would go home where I did nothing except sleep and cry. After about six months of this, I finally contacted a therapist and went on antidepressants. They helped some, but it took a good two years to get over the worst of it. I don't think I will ever feel the same.

Most of us experience depressive symptoms at some point in our lives, with as many as one in four women and one in seven men having a depressive "episode."[5] These episodes can last months, even years. People who are depressed often lose interest in doing the things that they used to love. This feeling—referred to as *anhedonia*—is a loss of pleasure from activities in life that usually give us pleasure, whether it be food or sex or exercise or going to a movie. Sometimes, thoughts of guilt, shame, and helplessness perpetuate the feelings of despair. But mostly, people who are depressed complain that they do not have energy. In his book *The Noonday Demon: An Atlas of Depression,* writer Andrew Solomon says, "The opposite of depression is not happiness but vitality."[6] And studies support his statement.[7] For example, people who are depressed tend to walk slowly and talk slowly. It becomes difficult to concentrate, to work, to read—basically to do anything.

Suffering through depression is tough, but doing so while going through a life trauma can be devastating. Some people never recover. Others do and almost don't recognize themselves afterward. I once talked with a woman who was depressed throughout her childhood and then became even more depressed after her mother died unexpectedly. She would lie in bed all day with her door shut; her children were left to fend for themselves. She knew that what she was doing was wrong, but she simply couldn't do anything about it. When she finally found relief with therapy and the clouds broke, she was surprised to realize that not everyone felt the way she had. She had been depressed for so long—basically her whole life—that she couldn't even imagine others not feeling depressed, much less imagine them being happy.

PSYCHOSIS IN THE AFTERMATH OF TRAUMA

Most of us know how stressful life events make us feel—depression, anxiety, and nagging thoughts about what happened or what could have been. These states of mind, disturbing as they are, do not necessarily prevent us from living our "normal" lives. But sometimes the stress of trauma is too much to bear, and functioning becomes impossible. A more extreme response is psychosis, which is a "break with reality." Contrary to what many people think, psychosis is not a diagnosis but rather a broad term used to describe a symptom or set of symptoms that can become part of a diagnosis. For example, people with schizophrenia sometimes experience psychosis in the form of hallucinations, which are changes in perceptions—such as hearing voices or feeling things crawling all over the body. They can also have delusions, which are distortions in beliefs, such as thinking someone is out to get them or that the person on the television is talking just to them. In the extreme, trauma itself can induce psychosis.

For example, *conversion disorder* is an extremely rare condition that illustrates just how far the brain can go to avoid thinking about trauma. One summer, I went to Cambodia with my son. Scattered across hundreds of fields—known as the Killing Fields—are the remains of more than a million people—men, women, children, babies—who were butchered to death in the 1970s. Standing there in one of the fields, we could see the bones and skulls and clothing still coming up from deep in the ground. It is hard to imagine how anyone could survive such a traumatic ordeal. And yet some people did. A few found themselves in ophthalmology offices—complaining that they could no longer see.[8] As expected, they suffered with PTSD, depression, and anxiety. But why couldn't they see? Today their diagnosis would likely fall under the general category of *somatic symptom and related disorders. Soma* means body. In this case, the body responded to the trauma with blindness as a means of breaking from reality. It's as if they could not bear to revisit the memories of what had happened to them.

MENTAL ESCAPE FROM TRAUMA

You have probably heard of multiple personality disorder, now called *dissociative identity disorder* (DID). This is a serious condition, often portrayed in movies and TV as an unpredictable person with unpredictable personalities. One of the most popular depictions was in *Sybil,* a television movie where Sally Field portrayed a young woman jumping from one personality to the next. In reality, people with this condition do not have different personalities. Rather, they have different *alters,* which are different expressions of identities within one body and brain. They can have different names and behaviors and even memories. Some may smoke; others do not. Some are right-handed; others are lefties. They can also be of different genders. I spoke with a therapist once who was treating someone with twelve different alters, one of whom wanted to transition. That said, many of the behavioral differences between alters are more subtle, suggesting a lack of continuity of self.

This syndrome is both shocking and fascinating to people. When I talk about it in class, the questions just keep coming—How does one alter communicate with another? How do they keep track of their memories? What happens when they are sleeping? The answer is usually, We just don't know. Like somatic symptom disorder, DID is rare; therefore, there are very few studies on it, and those that do exist are controversial.[9] What we do know is that most people with DID have experienced extreme trauma. One woman was severely abused by her father when she was a child. After dinner, he would tell her to take a shower, and as she stood naked in the shower, she could hear him coming up the stairs. She hated it, obviously, but could do nothing. So, as soon as she heard him coming up the stairs, she would escape internally and only return after she was alone and "safely" in bed. By the time she was an adult, she had so many identities that she could no longer function in everyday life. She needed help, which she did get—but only after a lifetime of suffering.

It is easy to understand *why* someone might want to escape into another identity, especially when traumatized as a child, but *how* it happens in the brain is more difficult to explain. One study examined the brains of women with DID or Dissociative Amnesia, which includes severe loss of memory. All of the women had been traumatized during childhood.[10] When the researchers measured the volume of certain brain regions, especially those involved in stress and trauma, they were less compromised than those regions in women who were also traumatized but did not have different identities. The researchers suggested that escaping into another identity somehow protected a person's brain from being damaged. This was a very small study, and the participants were also diagnosed with a personality disorder. As such, the interpretations of these results are not universally accepted.[11] But they are thought-provoking nonetheless. I present them to underscore the lengths that the brain can go to escape trauma—almost as far as becoming another person.

MOST OF US

In this chapter, I wanted to review the possible responses to trauma—even going as far as a break with reality. But to be clear, most people do not experience all or even some of these responses, and the majority would not be diagnosed with PTSD. But that does not mean that we do not feel the aftereffects of trauma or that we don't think about what happened to us or blame ourselves for what happened day after day. It is the thoughts and memories that keep our past experiences alive—both good and bad. So now, I want to dive deeper into these mental processes and why the brain makes them last so long—because if we can begin to understand why everyday traumas change our brains, we can begin to put the thoughts and memories it generates into perspective, at least that is my hope and my intention going forward.

3

The Two Forms of Everyday Trauma

The trauma was life-changing. At the time, I thought it was the worst, but it ended up being for the better. Every part of me changed after this day. I am thankful I went through it because I am now more aware that in life anything can happen. I just wish I did not have to go through the pain to get where I am today.

—Maggie, describing the emotional impact
of Hurricane Sandy

So far, I have been using the term *everyday trauma* to describe stressful events that happen suddenly and linger as thoughts and memories as well as traumatic experiences that continue day after day. Now I want to make a finer distinction between two forms of everyday trauma and the way the brain responds to each of them. The first one I refer to as *fast and fearful;* the response comes on quickly and generates significant feelings of fear. It could be caused by the unexpected death of a loved one, an earthquake, or a violent attack on a city street. In this case, the event produces a fear response, which tends to dissipate with time. I told you about my experience in the Hollywood Hills when a man tried to break into the house while I was sleeping. My fear that night was intense, but it quickly dissipated once I escaped and found safety. Occasionally I think about what happened that night, but for the most part, I don't. For others, acute experiences live on in their thoughts, which can elicit feelings similar to those experienced when the trauma occurred, almost as if the trauma were happening all over again. Take Ava, who had a close call when her infant quit breathing in the crib. She said that during the event, every part of her being was in a state of high anxiety:

> My mind was racing and I couldn't make sense of my thoughts, yet I was called upon to make lifesaving decisions. My heart was beating so fast that I thought it would stop from exhaustion. I just paced like a wild animal in a cage. Every part of me wanted to jump out of my own skin. I wanted to walk until I dropped from exhaustion so I wouldn't have to think anymore. To this day, when I hear the sound of a heart monitor or just think about what happened, I begin to cry.

Clearly, this experience induced fear and panic at the time and continues to be an everyday trauma for Ava.

The second form of everyday trauma emerges more slowly and is more persistent. I call it *slow and stressful.* The global coronavirus pandemic is a good example. In the early part of 2020, most of us were living our "normal" lives with our normal worries and concerns, as well as trying to accomplish daily goals and plan for the future. And then poof, just like that, it all stopped. Many of us were forced to leave our jobs and work from home; some of us lost jobs altogether. We became afraid of getting the illness or of our loved ones getting it or, even worse, dying from it. First responders were obligated to go out into the world to take care of the sick and dying. We became afraid to do the most mundane things in life. Just going to the grocery store required preparation, not to mention trying to deal with the anxiety it generated. We could barely imagine going on vacation or to the movies or even just out to dinner. And while living through it, we had no idea when it might end. It could be months or even years. We couldn't help but think that things might never get back to normal. We were stuck living with the everyday trauma of this new and unexpected reality. In psychology, the pandemic would be called a chronic stressor because of its persistence over time. It is also an everyday trauma, because it affected us—traumatized us—every day, regardless of the day's specific events.

The distinction between these two forms of trauma—a fast, fearful response and a longer, steadier one—is admittedly arbitrary. Most of our traumatic experiences contain elements of both. But intuitively, we understand the difference between them, in part because they are encoded in our bodies in distinctive ways. Indeed, we have two biological systems that respond to these two forms of everyday trauma.

THE SLOW STATE OF STRESS

Let me explain the slower system first, starting with a concrete example. Over the past few years, tens of millions of people have been forced to migrate because of war and famine. Along the way, they suffer all manners of stress and trauma—from family separation to danger, discrimination, and disease. And they are always living with uncertainty. Austria's Ricarda Mewes and her colleagues wanted to better understand the trauma associated with migration.[1] To do so, they measured cortisol, which is the major stress hormone in our bodies. This hormone is made in the adrenal glands (or adrenals), which sit on top of our kidneys. These tiny organs release cortisol into the blood when we are stressed, and from there cortisol goes everywhere in the body, including to the brain.

The researchers did not measure cortisol in the blood—this would be impractical given their need to look at patterns over extended periods of time. Rather ingeniously, they looked inside the hair. Hair grows very slowly, about an inch every two or three months (about one centimeter per month), and therefore, cortisol levels at a given point on a strand of hair can provide a timeline of stress. If the amount near the scalp is high, the person has likely been experiencing a lot of stress day after day in recent months. If it is low, the person has likely not been as stressed. The concentration of cortisol in the hair root can be thought of as a good indicator of cortisol levels in the brain. The researchers monitored three groups. One group had migrated from the Middle East, primarily from Iran and Syria, and had made their way to Germany, where they were seeking asylum. Another group had also migrated but were permanently settled in Germany as immigrants. And a third group comprised residents of Germany who had never migrated. The immigrants who were still seeking asylum had very high levels of cortisol in their hair, nearly twice as high as those who had successfully immigrated years ago. Interestingly, there was no difference in cortisol between asylum seekers with and without PTSD. Moreover, cortisol levels in people who had never migrated (residents) fell somewhere between the other two groups'. Thus, the immigrants who found refuge were releasing less cortisol into their blood and eventually into their hair.[2] These data, albeit interesting, tell us what we logically already know. Seeking asylum is stressful and traumatic. Finding refuge is a huge relief to the body—including the brain.

How do the adrenals know they need to release cortisol? The signal comes from the brain—deep within the brain—from a tiny structure called the *hypothalamus*. When something bad is happening or is about to happen, the hypothalamus sends out a molecular messenger into the body, which eventually causes the adrenals to release cortisol into the blood—which then goes back to the brain to send out yet another messenger that shuts off the release of cortisol. This system is self-regulating—kind of like the thermostat in your house. You set your desired temperature. When the thermostat turns on, the furnace increases the release of warm air, and when the right temperature is reached, the furnace turns off. Your stress system works in a similar way because it responds to change. By doing so, it keeps your brain and body working in the right zone, especially when the unexpected happens.

STRESS HORMONES CHANGE THE BRAIN

Extremely stressful life experiences often stay with us for the rest of our lives, and even small stressors remain on our minds. The brain changes as a result of these life events, but for most of modern history, neuroscientists did not know how it happened. Then in the 1960s, Rockefeller University's Bruce McEwen discovered that when cortisol enters the brain, it remains there by attaching itself to a receptor.[3] Receptors are like locks in a door, and this particular receptor has a molecular structure that can only be opened by cortisol. McEwen found these receptors all over the brain, including in the hippocampus, a part of the brain that is especially sensitive to stress and is also used for making new memories, including those about our most traumatic life events.

These results were exciting because they suggested that hormones released during stressful times might change the actual structure of the brain. A few decades later, two enterprising scientists working in his lab, Catherine Woolley and Elizabeth Gould, decided to test this idea by giving rats daily injections of a relatively large amount of corticosterone, which is the rat version of cortisol in humans. Their experimental protocol was meant to mimic a situation in which the brain is exposed to a large amount of stress day after day, like the migrants who were trying to find asylum or the many people living in fear and striving to survive through the pandemic. Then they examined the anatomy of the brain under the microscope, focusing in on the connections between neurons, known as *dendrites.* These connections look like a tree with many branches splitting off in different directions. The scientists noticed that the neurons that had been exposed to the stress hormone had fewer branches. Some of the neurons even appeared "shrunken."[4] These discoveries, and similar reports, uncovered a mechanism whereby stress can change structures in the brain. Moreover, they suggested that extreme stress, if experienced chronically over time, was not good for maintaining the brain's integrity and, presumably, its ability to function properly.

But not all stress is bad for the brain, and as a result, the presence of stress hormones is not necessarily bad, especially in short, quick doses. For example, we did a study during which male rats were injected with just one large dose of corticosterone (again, the rat version of cortisol), and they learned better.[5] Similarly, if they were exposed to a short, stressful event that would cause a similar release of the hormone, they learned better. And importantly, this effect depended on the release of hormones from the adrenal glands, meaning that the rats did not learn better if the adrenal glands were removed. This increase in learning was accompanied by anatomical changes in the brain—again in the hippocampus. As just described, neurons are connected to one another with dendrites. On these dendrites are tiny protrusions, called *spines,* that sprout out from the branches to connect one neuron to another. We counted those spines—not easy to do. And the male rats that learned better had more of them after the stressful event. It's as if their brains were getting ready to learn—and then they did.[6]

So, do stress hormones affect the human brain in a similar manner to rats? And perhaps more important, can stress hormones affect what we think about and remember in our everyday lives? In one study, blood cortisol levels in more than two thousand people were measured. Then the people were asked to complete a battery of cognitive tasks. During these kinds of studies, a person is typically read a story and then asked to repeat as many details as they can remember or shown pictures of objects that are cut into pieces, which they are then asked to identify.[7] Overall, the people with the most cortisol in their blood were less likely to remember the details of the story and performed less well on the other related measures of cognition. Obviously, the scientists could not look at connections between neurons under a microscope (this can only be done in postmortem brains), and so instead they used magnetic resonance imaging, or MRI. This technique images the whole brain and then estimates its volume. The people with more cortisol in their blood had less brain volume in numerous regions. In particular, they had less gray matter volume, which indicates the presence of fewer neurons. But interestingly, this effect was observed only in the women. Another study also tried to associate cortisol with memory, this time in the saliva. The people were read a story and then exposed to an acute stressor, during which they had to place a hand in very cold water. One week later, participants were asked to recall the details of the story. These results were a bit different from the last one. This time, the participants who released more cortisol during the stressor remembered the story better—at least the men did. Cortisol levels did not relate to memory for the women.[8]

It might seem like the results from these studies are somewhat contradictory, with cortisol impairing memory in one case and enhancing it in another, not to mention differing responses between men and women. But remember that the cortisol levels and memory were measured at completely different times and under different circumstances. In the first study, cortisol was measured in the morning before the experiment even started, which would reflect its baseline level, and there was no real stressor, whereas in the second study, the hormones were being released in response to the brief but painful feeling of the very cold water—an acute stressor. Therefore, we can't measure someone's hormone levels at one point in time and come to sweeping conclusions about how one stressful event will affect one person's ability to learn and remember, or even how it will make them feel.[9] What we can glean from these data, and the thousands of other studies on stress and the brain, is that stress hormones are always changing and they affect the way we think and learn and remember, depending on when and how much is released over time. Whether these changes are meaningful to everyday life depends on the individual—their age, gender, sex, life experiences, and even how resilient they are.

TRAUMA FROM ONE GENERATION TO THE NEXT

In some cases, the impact of trauma can stretch beyond one lifetime and even persist over multiple generations. Mother rats take care of their offspring by putting them together in a pile and crouching over them. Underneath, the baby rats latch on to her nipples to get milk while being kept warm and safe from predators, before they learn to roam around and survive on their own. In the 1960s, Stanford neuroscientist Seymour Levine observed that while under the mother, the offspring were developing their own stress responses. In other words, their adrenal glands were "learning" to release corticosterone into the blood and thus the brain, which in turn was learning to send a signal back to the adrenals to stop releasing the hormone. This process, once fully developed, allows the animal to respond to stress as an adult. However, Levine noticed that when the the young rats were not touched or cared for by their mothers, the stress responses did not develop normally. In his studies, offspring that were separated from their mothers for just a few hours a day released more stress hormone in response to stimuli in the environment when compared to the amount released in rats that had not been separated from their mothers or were gently handled, mimicking what their mother would have done to them.[10] In general, these changes in the stress response seem to be induced by the behavior of the mother and not inherited through her genes (DNA). Mother rats that care for their offspring tend to produce offspring with a regulated stress response, regardless of whether they were the biological mother or a foster parent.

Intergenerational transmissions, where parental experience is transferred to offspring, are often called *epigenetic*, meaning that they occur outside of inherited genetics. Indeed, many of these transmissions are produced through changing the expression of RNA, which is used to make proteins, often referred to as the building blocks of the body, including the nervous system.[11] These transmissions can even begin before birth. In one study, researchers analyzed data from the womb into adulthood. The participants were categorized into those who were exposed to stress in utero and those who were not, and as middle-aged adults, their brain activity was assessed using a functional MRI machine. This measurement is considered *functional* because it measures how the brain is responding in real time based on the amount of blood being used by neurons to process information. The women who were exposed to more intense stress when they were fetuses had more brain activity, especially in the hippocampus, which, as I mentioned, is very sensitive to stress and stress hormones. Low levels of stress in both sexes were associated with more activity in the hypothalamus, which is involved in the stress response. These findings underscore long-term and sex-dependent changes in the brain as a result of stress while we are still developing in our mothers' bodies.[12]

Generational studies such as these are nearly impossible to carry out, and when they are, the analyses are more often than not retrospective. Scientists cannot manipulate variables ahead of time (as they can in laboratory studies) but must rather rely on written or other types of recorded information. Regardless, some of the findings are intriguing and, as a group, persuasive. One of the most famous studies of this kind describes the Irish potato famine that occurred in the 1800s, during which a million people were killed, with another few million migrating to avoid almost certain starvation. Retrospective analyses indicated that the people who at the time of the famine were still developing in their mothers' wombs were more likely to be institutionalized with mental illness as adults.[13] Apparently, maternal responses to the psychological and nutritional stress were somehow passed on to the fetus and then expressed within offspring as a change in overall mental health much later in life. More recently, neuroscientist Rachel Yehuda and her group documented an increase in stress-related symptoms in the offspring of mothers who were pregnant and living in New York City during the 9/11 attack.[14]

As you can see, experiences of stress and trauma not only change our own thoughts and feelings but may well extend into the health and well-being of others, even our children and their children's children.[15] The power of what we inherit both genetically and epigenetically does not relate only to trauma, though. We also have the ability to inherit well-being and positive thoughts and feelings. In my line of work, we are often focusing on the bad, but the epigenetic influences of our parents and grandparents also bring hope and wellness to our lives.

THE FAST FEELING OF FEAR

Now let us consider the other form of everyday trauma—the fast one that induces fear. This is the type of trauma that most people think of when they think of *trauma*. The response is designed to make us spring into action in the moment. And it is designed to make the brain quickly record memories of what is happening while it is happening. In 2012, many people were affected by Hurricane Sandy, among the deadliest storms ever to hit the East Coast. Maggie, who lived in a house near the ocean, was warned about the storm and knew it could be bad but had no idea how bad. Once the storm hit, its severity began to dawn on her. She remembered:

> My body went into full-blown fight-or-flight mode. My actions were fueled purely by instinct and adrenaline. The events of the night became blurry after the power went out and we were still trying to bail out the water. I do very distinctly remember the moment of accepting defeat, though. The crushing, sickening feeling of helplessness. The fear of the unknown danger and destruction happening all around us. I felt my heart racing and my throat choked up from holding back tears, but I also remember feeling virtually numb during these robotic moments. Thinking back on it, I suspect this numb feeling came from an overwhelming amount of stress and fear, but also shock.

Maggie describes a number of thoughts and feelings happening all at once. She is afraid; she is busy; she is frozen; she is about to cry; she is numb; she may even be in shock. Her whole body is involved. So, when we consider the feelings of fear, we have to include the whole nervous system—not just the brain. The brain is connected to every other organ and part of the body through nerves, which are like very long roads through which electrical signals flow back and forth every second of every day. This system of nerves is called the *autonomic nervous system* (ANS) because, as its name implies, it is more or less automatic. The ANS controls our breathing, keeps our hearts beating, and prompts us to sleep. It is absolutely necessary to have this kind of system in our bodies because if we had to "remember" to breathe or sleep or eat, we would not survive for long and we wouldn't have the conscious bandwidth to do everything else we need to do. Within this system are two subsystems—or arms. One arm is called the *sympathetic nervous system*. It is most often referred to as the *fight-or-flight* response because it becomes engaged when we need to either get away or stay and fight. It is activated when we are afraid and experiencing a trauma. The other arm is called the *parasympathetic nervous system*. It is used to get us back to baseline—to homeostasis—and is referred to as the *rest-and-relax* response.

Imagine, for a moment, that you are walking home alone on a shadowy street on a dark night. You suddenly become aware of a person walking up behind you and getting closer and closer. You start to panic. It is time to think quickly and make decisions. Should you stay and fight or run for your life? Immediately, your brain sends a signal to the adrenal glands. While releasing cortisol, the adrenals also release adrenaline, more commonly referred to as *epinephrine*, which is used in the body to increase the heart rate. Normally, we don't need a lot of epinephrine released into our bodies—just enough to keep us awake, thinking clearly, and behaving wisely. We release a bit more of it when we become aroused, still more when we need to escape, and even more as we become afraid. Epinephrine makes the heart beat faster in order to get more energy into the rest of the body—especially the brain. However, unlike cortisol, epinephrine cannot enter the brain on its own. Instead, it provides quick energy to the brain in the form of glucose.[16] It also uses some of the nerves in the spinal cord, such as the vagus nerve, to send messages to the brain that influence our decision-making and the memories we retain about what is happening.

Imagine once again that you are walking down a dark street alone when someone comes up quickly behind you. Your heart starts beating faster as your brain starts to think about what you should do next. Then you hear the person call out your name, and you think that you recognize his voice. You look and see that it is your neighbor. Right away, you breathe a sigh of relief. Your heart quits racing, and your body relaxes. No more danger! Whew! What is happening in your body now? The parasympathetic nervous system has kicked in. This system is mediated by another neurotransmitter known as *acetylcholine* and the vagus nerve, which again sends information to the brain. This time, the brain sends a signal back out into the body to calm everything down—when and if the danger has passed. Once this system has established itself, you might even get an urge to have a snack, have sex, or just relax and go to sleep.

BRAIN AND BODY WORKING TOGETHER

The sympathetic and parasympathetic arms of the ANS are involved in just about everything we do—from breathing to sex to surviving trauma.[17] As we breathe in, our sympathetic system is active, and as we breathe out, the parasympathetic kicks in. When we become sexually aroused, the parasympathetic is activated, but later, with orgasm, the sympathetic is more engaged. To stay alive and healthy and behave wisely, we need both arms to work well together. And we need both arms to integrate well with the other stress response—the slower response that uses cortisol to change the structure of the brain. A young man, Alex, tells the story of a traumatic experience in his life that illustrates all of these systems working together to help him survive and recover:

> The first time I was in a car accident, I don't think I realized I was in a real car accident until everything calmed down. I remember going with my best friend through a large intersection that we had crossed a million times before in our town and thinking nothing of it. However, when I raised my head, everything sort of slowed down. I watched a big red SUV come at my door, and it seemed like it was going super slow and some distance away. In the next instant, we were fifty feet away from where we had started, and my friend's car was missing the front end.

Alex recalled feeling calm immediately after the accident:

> I felt like I should have been crying, but I felt this immense sense of calm. Usually, I'm super stressed out as a person, so this was weird to me. My body felt full of energy—but not tingling like when you are excited. Just a simple "If you need me, I am here." The EMTs made me get onto a stretcher and into an ambulance just in case I was injured. I felt fine, if a little lost in my own thoughts until I was in the hospital and saw my mom crying. Then I realized that this was a scary situation.

Let's go over what happened in Alex's body while he was having all these thoughts and feelings. Within several hundred milliseconds, the visual cortex in his brain "saw" the other car coming for him and within several hundred more milliseconds predicted that it was going to hit their car. This information was quickly sent to his hypothalamus, which responded by activating the adrenal glands to release a big bolus of epinephrine. This caused his heart to beat faster, sending more oxygen to his muscles—and to his brain. He might have needed those muscles to make a quick escape or save his friend. And he definitely needed oxygen in his brain to make a quick decision about what to do next. Once he realized that he was safe and alive in the hospital with his mother, his heart stopped pounding, and he finally relaxed. Meanwhile, the cortisol that was released into the blood during the accident had time to distribute itself around the body to reduce inflammation and help it recover from injury. It had also reached the brain to help generate anatomical structures that will consolidate memories of the event, which will likely stay with him for the rest of his life. Indeed, Alex remembers most of what happened that day. He may not think about it often, but the memory is there, lurking in the background ready to be used in case something bad like this ever happens again.

As we come of age and then get older, we all are faced with challenging times and situations. Some are short and fearful, while others are long and stressful, but no less profound. When they arise, we have to be ready to learn from them, all of them. To do so, we need our brains to elicit both types of responses in our bodies: we need the fast response to generate fear and action in the moment, and we need the slower stressful response to help us recover and learn from what happened. And we need them to work well together, with our brains firmly in control.

PART II

From Thoughts to
Memories to Feelings

4

Ruminations: Thoughts
That Get Stuck in Our Brains

Was it traumatic? I guess it probably was even though I did not appreciate it at the time. From that point forward nothing about my life was ever normal again.

—Melanie

When I was in my twenties, I became friendly with a woman named Melanie who was about my age. We were hanging out in my room one day and sharing details about our lives, when she started telling me stories about the sexual abuse she'd suffered growing up. She told me about Bradley, a good friend of her dad's. He groomed her for months before initiating sex with her when she was only eleven. He was thirty-five. She said,

Bradley worked at seducing me. He bought me presents and told me I was pretty. Then he, very slowly, began to touch me more and more intimately, a little bit at a time. Finally, we had sex. I thought we were having a real relationship—even though it was a relationship I knew I couldn't talk about. I would stay at home waiting for him to come over while girls my age were off doing normal little girl things with each other. This went on for years. It was very strange and changed everything about the way I viewed the world.

I never asked my friend, but I assume that memories of these experiences run through her head, playing over and over again. These trauma-related thoughts are clinically referred to as *post-traumatic cognitions.* They tend to revolve around a specific event and go around and around in the brain. *This experience changed my life forever and I will never be the same.* Or *I keep thinking that something like this will happen again.* Or *It seems like I did something to make this event happen.* These types of thoughts are assessed with a survey known as the Posttraumatic Cognitions Inventory, or PTCI, which measures how often someone is thinking about traumatic life events.[1]

People with PTSD usually have many post-traumatic thoughts, so many that they have trouble functioning in everyday life. But people without PTSD have them, too. They are a "normal" response to trauma. The brain is primed to think about bad events that happened in the past so that we can remember to avoid similar environments or conditions or people in the future. In chapter 2, I mentioned a study that my group did with women who had experienced sexual trauma. Even though most were not diagnosed with PTSD, nearly all of them reported having many trauma-related thoughts. These thoughts, disturbing as they are, do not have to interfere with everyday life. Most of us can still get up in the morning and go to work or school or take care of our families and ourselves, even with these types of thoughts coming and going throughout the day. Nonetheless, these thoughts indicate that the trauma is not forgotten and is still impacting us negatively.

RUMINATING ON OUR LIVES

As time passes, most people who have experienced trauma have fewer and fewer of these kinds of thoughts—the ones that are specifically about a traumatic event. These thoughts may shift and mutate. The person may start to have thoughts that are more inwardly directed and even more repetitive. In some cases, we don't even realize these thoughts are related to the original trauma. In my friend Melanie's case, she may start to think about her own life and how she always does the wrong thing, or she may start to think that she is a bad person who can't change. When these thoughts cycle through the brain over and over again, they are called *ruminations*. A ruminating thought gets repeated again and again . . . and again. In fact, the verb *ruminate* comes from Latin, meaning "to chew." It was initially used to refer to cows, who chew their cud over and over again, even after it has lost its nutritional value—trying to get every last bit out of it. This is also true for a ruminating thought. The thought has lost its nutritional value and is not necessarily going to produce a solution, but our brains repeat it over and over anyway.

Ruminating thoughts are commonly mistaken for worries, but they are different. Worries have intention, whereas ruminations typically do not. Ruminations are autobiographical and directed at oneself. Worries also tend to be autobiographical, but they are usually focused on solving a particular problem or addressing a particular situation. Ruminations are focused primarily on one's own distress—not on the details of what happened or what caused it. For example, someone might be silently going over all the things that are wrong with their life without any real intention of changing anything. Interestingly, when you ask people why they ruminate, they usually tell you they are doing it to figure something out—they are trying to solve a problem. So, even though ruminations may seem useful to the person who is producing them, generally ruminating thoughts don't bring positive revelations or change in our lives. Ruminating keeps our attention on ourselves and weakens our ability to pay attention to what is happening in the present.

To be clear, you do not have to be traumatized to ruminate. When I was younger, I ruminated a lot. I would lie in bed at night going over and over things that had happened in my life, even writing them down in shorthand. Night after night, I would go over the memories again, using my notes as cues. If you had asked me, I would have told you that I was trying to figure out what happened and why it happened—trying to construct a coherent story. Looking back, I think I rather enjoyed it, almost like a game of detective. But even though I was attracted to these kinds of thoughts, they were not helping me. Instead, I was just becoming more anxious and distracted. I am still a ruminator to this day; it's a hard habit to break! Of course, I am not unusual; we all ruminate. As clinical psychologist Edward Selby sees it, ruminating is like kicking a soccer ball around the field with no game plan.[2] And thus there are excellent reasons for limiting the time we spend ruminating. But to do this, we must first become more aware of when we are doing it—and why.

People have been generating ruminative thoughts since long before psychologists began studying them. But one psychologist in particular, Dr. Susan Nolen-Hoeksema, began to study them in the context of mental health and, as a result, put them on the map, so to speak. She died tragically in her early fifties but left behind a legacy of research on rumination, including the so-called RRS, which stands for Ruminative Response Scale.[3] This questionnaire is used to find out exactly what people are ruminating about and how often. There are several versions of the RRS, but in general, someone is presented with statements similar to these:

- I think about what I said or did in the past few days.
- I think about why I can't focus on one thing at a time.
- I think about why I always do the same things over and over again.
- I think about how alone I am.

The person then ranks each statement from one to four, with one meaning "almost never" and four "almost always." A higher score is indicative of more rumination. My research team has been using this survey for years and in general, we find what others find: people tend to ruminate more when they are stressed and/or depressed; they tend to ruminate if they have been traumatized; and women seem especially inclined to ruminate.[4] This might even explain in part why women are more likely to be diagnosed with stress-related problems such as PTSD and depression (more on this in chapter 6). But there is at least one piece of good news: people tend to ruminate less as they get older.[5]

WHY RUMINATE?

If ruminations are so bad, why in the world do we engage in them in the first place? The distinguished psychologist Caroline Blanchard from the University of Hawaii speculates that ruminative thoughts are actually quite useful under the right circumstances.[6] She suggests that they are used to assess risk—something called *risk assessment*. In every moment of our lives, we are assessing risk: *Is this situation dangerous? Am I safe? Who will I meet or see here? What might happen next? Can I escape if something bad happens to me?* As we go about living our lives, situations arise that are ambiguous. For example, if someone breaks down a door in your house while you are sleeping, the risk is not ambiguous. You know immediately that you need to respond. You need to act and do it fast, even if it involves freezing. But if you wake up in the middle of the night and hear a sound from somewhere outside, this is more ambiguous. Is it really something to worry about, or is it just a stray animal walking around? Is your neighbor just having a late-night party? Imagine what you would do if you are walking down the street and a man you knew pretty well, but not super well, comes up from behind and grabs you at the waist. Maybe he is just fooling around. Maybe not. Maybe he has had too much to drink and is trying to attack you. Perhaps you are especially worried because you were attacked on a street like this before. Perhaps someone robbed you on this very street in the not-so-distant past.

Most situations in life carry some level of risk, even if only the potential for embarrassment. We are always on guard for it. And the brain is usually ready to respond. But it does need time to think—*What should I do?* It might need some extra time to retrieve a memory from the past—something that helps us decide what we should do. According to Dr. Blanchard, we need time to assess the risk. If we think of ruminations in this way, then we can imagine that they have value—immense value. Perhaps they evolved to help us assess risk in the moment and to make quick decisions. Thinking of them in this way, ruminations probably save lives and are therefore good for us. But like many things that are good for us, they are only good in moderation.

RUMINATING ON OURSELVES

Ruminations are not always about the past, but because they are repetitive, they always engage processes of memory. I like to think of them as memories laced with mood. Some psychologists distinguish between three types of ruminations. There are depressive ones, brooding ones, and reflective ones. The first type, depressive ruminations, are thoughts that occur when someone is thinking a lot about how sad they feel and how alone they are. To date, most rumination studies have focused on people diagnosed with or experiencing symptoms of depression. This is because ruminations are prevalent in people who are depressed. In fact, depressive ruminations are difficult to distinguish from symptoms of depression—*Why am I so lonely? Why am I so sad? Why can't I seem to concentrate?* However, while depressive symptoms are generally "felt" in the body, ruminations are thoughts.

They are nonetheless tightly linked, as are many of our thoughts and feelings. In general, people who are depressed tend to have more of these thoughts, and conversely having more of these thoughts can exacerbate feelings of depression. In one of our studies, we reported a correlation of 0.7 between depressive symptoms and depressive ruminations.[7] A correlation of 1.00 is perfect, with no measurable difference between two variables. Therefore, 0.7 is a meaningful correlation.

The second type of ruminations, the brooding type, are even more self-focused than depressive ones. They tend to involve blame—usually self-blame. An example of this is, *Why do I always make the same mistakes—over and over again?* or *Why can't things ever go right for me?* or *Why can't I learn to handle my life better?* or *Why doesn't my partner understand what I am experiencing?* Brooding thoughts tend to be passive, with no serious intention to change things or solve problems. They are especially prevalent in people who believe they do not have much control over their lives.

The third type of ruminations—the reflective ones—are about the past: *What happened and why did it make me feel so bad?* or *What can I do about what happened to make things better?* or *I wish I could just figure out what is going on.* These thoughts require more concentration. Sometimes they are accompanied by deliberate action, like the note-taking I did as a young adult, lying in bed at night thinking about what happened in my life. Reflective ruminations are also more analytical than the depressive and brooding types, and may even be more adaptive. They are considered more intentional, as if pondering. In fact, some researchers think that reflective ruminations should not be grouped together with depressive and brooding ones because they are so different in content.[8] But regardless of their content, these kinds of thoughts still take attention away from the present moment, without necessarily helping to solve problems of the past or future.[9]

HOW RUMINATIONS DISRUPT
OUR LIVES—AND OUR BRAINS

On the face of it, it might not seem so bad to ruminate, especially if you are just assessing risk or trying to figure out what happened in the past. But while ruminating, we aren't doing other things. Ruminations keep us from paying attention to what is happening now—at this moment. In one study, people who were depressed and ruminate frequently were asked to respond as fast as they could to a probe flashing on a computer screen after various words were presented.[10] They responded similarly to controls when neutral and positive words were presented. However, when the participants were shown a negative word like *incompetent* or *worthless* or *flawed,* their attention was captured by and biased toward the negative information, especially in participants who were ruminating the most. In another study, people were asked to engage in a very demanding task by keeping a long list of words in memory over and over again.[11] They were then asked to read short stories with emotional and negative content. The introduction of this information reduced their ability to keep the words in mind, especially in those who were ruminating the most, again suggesting that ruminations interfere with our ability to use our brains well in the present moment. These kinds of laboratory studies are artificial and do not necessarily reflect how we would respond in our everyday lives, but I can imagine being one of those people!

What about the brain? What is it doing while we are ruminating? How do ruminations interfere with its ability to concentrate? My Rutgers colleague Brandon Alderman and I did a study in which we asked people who ruminated often to concentrate while we measured activity in their brains.[12] Participants were instructed to focus their attention on an arrow on a computer screen and respond by selecting the correct direction of the arrow: up, down, right, left. It is a simple task to do. But then we added in other arrows pointing in different directions all around the target arrow, making it more difficult for participants to control their attention. As such, this measure is often used to assess "cognitive control." Then we looked at their brain activity during the task. Normally, the target arrow is "seen" by the brain, and as a result, neurons fire all at once. This kind of brain activity is called *synchronized* because many neurons are responding at the same time, and in general you want this kind of brain activity. It is how information gets passed around from one region to the next. However, the participants who reported the most ruminations had less of this activity—fewer neurons were firing at the same time in response to the target.

It is hard to know exactly what these data mean. Minimally, they suggest that the ruminating brain is not responding as it normally would or should and that, moreover, information from the outside world is not getting into the brain as efficiently as it could. This may not seem like a big deal, but it would be if you were supposed to be paying attention to something or someone you really care about. I heard a tragic story about a parent who was ruminating while sitting on the beach as the children swam in the rough ocean. Or consider an employee who was not paying attention and sent an incriminating email to the whole company. We all become lost in our thoughts from time to time, but once we become lost in ruminations, problems will arise. Our brains need us to pay attention to what is happening now.

RUMINATION AS A PROXY FOR MENTAL HEALTH

Symptoms of stress and trauma oftentimes coalesce within a person. In other words, if one person is experiencing one symptom, such as depression, she is most likely experiencing another one, such as anxiety, and perhaps another, such as rumination. If one person is thinking often about a past trauma, they are likely to be feeling stressed out in their everyday life. And conversely, if another person is not expressing one symptom, she is less likely to express another symptom or feeling. If she isn't especially stressed, she may be less likely to think about past traumas or be depressed. For example, my research team has been working with a group of women with HIV, many of whom were infected when they were born or as young children. One woman contracted HIV from her mother during childbirth, after which her mother died. She was passed around from foster home to foster home, sometimes forced to eat alone in her bedroom with plastic utensils because her foster family was afraid that she would infect them. And now years later, she lives with the memories of what happened to her. And along with those memories come feelings of sadness and anxiety about the future. Another woman had so many traumas in her life that she couldn't even remember them all. But she does realize how these traumas make her feel today—stressed, depressed, anxious, and worried about the days to come. Despite all their trauma, many of the women are tough and resilient. Others seem to struggle and suffer more, telling us that they could not quit thinking about the past and what had happened to them along the way.

As discussed in chapter 1, differences among people are called *individual differences,* and they are important. They tell us that stressful life events are not all equally traumatic and that any one person's response to them can be unique. And the differences exist for rumination as well. Some of the women in our studies said they ruminated often and others not as much. Moreover, those who ruminated the most were also those who were most likely to say that they were anxious and depressed and feeling stress in their day-to-day lives. And if they were ruminating often, they were much more likely to be thinking about traumas from the past. So, some questions arise: Is there a common underlying construct? Is there a common "factor" that can help explain why these various feelings and thoughts and symptoms are arising together within a person? To test this idea, my then graduate student Emma Millon entered the data from all the various questionnaires for all the various thoughts and feelings into a factor analysis. The surveys are common ones used in psychology to estimate symptoms of stress and trauma, such as Beck's depression and anxiety inventories (a survey for perceived stress) and the RRS scale that I mentioned for ruminative thoughts. We also included data from a scale that assesses "introspective awareness," which attempts to describe the thoughts and feelings that someone has about their own body and what their body is telling them about how they feel.[13] They can be estimated with questions such as: "When I am tense, I notice where the tension is located in my body" or "I notice how my body changes when I am angry." With all this information, the analyses then identified a common factor that we call *mental health.* This factor is considered the principal factor because it accounted for most of the variance—in this case 66 percent—meaning the majority of the differences. And in general, the women who endorsed the most maladaptive symptoms, such as anxiety, depression, and trauma-related and ruminative thoughts, also felt as if they could not trust their bodily sensations when compared to women who were experiencing fewer of these symptoms.[14]

You might not find this surprising because many of the questions on these surveys are similar, not to mention the fact that thoughts and feelings we experience do not exist in a vacuum but rather overlap to a great extent within our own brains. But then we advanced a more compelling question—what does this general factor of mental health predict? And of all the various outcomes that it could have predicted, it most predicted rumination, accounting for 94 percent of the changes in rumination, again among all the participants. In statistical terms, this is a huge effect. Now, to be clear, this was a small study and the other outcomes (anxiety, depression, stress, etc.) were also highly predicted by the overall measure of mental health. But still, rumination seems to represent and may even serve as a proxy for overall mental health and wellness—although not in a good way.

MAKING MORE MEMORIES

But why? Why are ruminative thoughts seemingly so bad for our mental health? Remember, ruminations are repetitive. And, there-fore, while ruminating, we are generating memories. But not only that, each time we have a thought about an event, we create a new version of that memory. And engaging in this process, especially if done over and over again, generates more and more memories, which has consequences. First, let me explain a bit about memory. We have memories all over our brains for various things that have happened in our lives, some of which we may not even think we can remember anymore. But as long as the memory was laid down in the first place, the memory is likely still there in the brain. When we go to retrieve a memory, this engages yet another process—retrieval. This process lets us visit the memory almost as if we are reliving it. We might see it in our minds or hear familiar sounds or have a familiar feeling. Imagine bringing a memory back from long ago. Now that you have brought that memory back into con-scious awareness, you are experiencing it in a new time and place: now. And accordingly your brain makes a new memory of what is happening now.[15] The memory of now gets attached to the old memory that you were revisiting in the moment—*What are you doing? Where are you? Who are you with? What time is it? What day and year is it? How do you feel? Do you like this memory any more now than you did before?* Some of this new information gets linked up with the old memory as it becomes updated. Now you have a new memory, which then gets stored in the brain along with parts of the old memory.

Let's think about how these processes might work in someone who has experienced trauma and is ruminating about what happened in the past. With each rumination, the person is making more and more memories—each time they ruminate about something from the past, they bring an old memory into the present moment. In this way, a traumatic memory becomes associated with new information, including how someone might feel about their own self now. As a result, this person has even more memories of the trauma and more associated thoughts and feelings about who they are in relation to the trauma—sometimes along with feelings of depression or blame or regret. Now to be clear, thinking about our past traumas every now and then is probably not going to have a huge impact on the brain. The brain is fairly resilient in this regard. But if we are engaging in these kinds of activities day in and day out for days, weeks, and years, we are adding memories on top of memories. Therefore, the trauma becomes represented by more than just the memory of what happened, because now we've added all the new memories that relate back to it each time it is rehearsed. Theoretically at least, ruminations might very well be filling our brains with more and more of those unwanted memories.

Think of your brain as a hard drive on a computer. If you save a document under a new name every time you make a change, eventually you will have so many copies of the document that your hard drive will fill up. In the case of your computer, you can do something about it. You can delete some of those old files or get a new computer entirely or store your data in a cloud somewhere. Sadly, we cannot simply delete old traumatic memories from the brain, much less get an altogether new brain. Our memories stay with us, most of them for the rest of our lives. Here is an example. Nora remembers one particular day as a member of a rescue squad. The ambulance was called for a woman in cardiac arrest. When she arrived, the woman was facedown on the bathroom floor, surrounded by a pool of blood. She tried CPR, but the woman was pronounced dead before the ambulance made it to the hospital:

> For months and months, I could not think of anything else. Whenever I closed my eyes to sleep, I would see the woman on her bathroom floor. The other members of the squad explained to

me that death was a part of the job, but even so, I still remember everything about that night and think about it often, even though several years have passed. I keep wondering if I could have done more to help this woman survive. And every time we get a new emergency call, the image comes back to me in my head. I still want to help others but just don't know if I can handle the stress of watching others die and revisiting the memories in my brain. Mostly I wonder whether I have the skills to respond in a crisis, even though this is what I am trained to do and it is my life's passion.

Again, it is entirely normal to have traumatic thoughts after a traumatic experience, and it is normal to worry about the future and to ruminate to some extent about what has happened to us in the past. But it is also okay to begin to recognize these thoughts as they arise and notice when they have begun to interfere with our normal everyday lives.

5

The Brain Is Always Learning

Life is all memory, except for the one present moment that
goes by you so quickly you hardly catch it going.

—Tennessee Williams, from
The Milk Train Doesn't Stop Here Anymore

In ancient times, scholars believed that mental processes like thoughts
and memories were generated within the heart. In fact, Aristotle
(384–322 B.C.) thought the brain was used to cool down the warm
blood coming out of the heart. This was a reasonable hypothesis,
given the available science at the time. When you have a frightening
experience or even just a bad thought, you can feel your heart beat
faster, but you can't feel anything happening in your brain. Now, cen-
turies later, we know that the brain makes memories and re-creates
them for our rehearsal, editing and updating them along the way.
And we know that memories are generated through activity within
specialized cells called *neurons,* which are unique to the central (i.e.,
brain) and peripheral nervous system. These cells are special. Without them, you wouldn't be able to think, move, or even think about
moving. You wouldn't be the person you are. In fact, you wouldn't
even remember who you are.

The human brain contains billions of neurons, and most of them are similar in structure. They each have a nucleus containing genetic materials (DNA), which are transcribed into RNA and then translated into protein (the building blocks of the nervous system). I like to think of neurons as houses in a suburban neighborhood. Many houses resemble each other from the outside. They often have a similar framework and structure, but each is a bit different depending on how and when the house was built and what color it is painted. The inside is especially distinctive because it depends on what kind of pictures hang on the wall, which appliances are being used in the kitchen, who has lived there before, who lives there now, and what they are doing in their everyday life. In short, each neuron is different from any other neuron depending on where it is located in the brain, what has happened to it in the past, and what is happening inside it right now.

Neurons do not work in isolation but rather depend on communication with other neurons throughout the brain to translate and transfer information. To do so, they each have a fiber that takes information away from the cell body, where it forms a "synapse" (a type of electrochemical connection) onto the next cell. Continuing with the neighborhood analogy, neurons communicate with one another like houses connected along a small path. Some paths are frequently traveled and have lots of activity. These pathways become more like roads, perhaps even paved roads. Some roads are very popular, with cars flying by. The most highly traveled roads become highways— faster and more efficient. This idea, that connections between neurons become bigger or faster or stronger with experience, is often referred to as *neuroplasticity*. Consider synaptic spines, those tiny protrusions that make connections between neurons. In one of our studies, we observed more of them after learning had occurred.[1] This is a good example of neuroplasticity because the structures arise rapidly in response to a new learning experience.

But anatomy, in and of itself, is not sufficient to produce a memory. Rather, the brain uses its specialized anatomy to produce a phenomenon somewhat like electricity, called *electrophysiology*. The current is generated by ions—mostly potassium and sodium—which carry an electrical charge back and forth across the cell membrane. Neurons generate either more or less current and thus become more or less "excited," but they are always generating this type of activity. Electrophysiology is how neurons communicate with one another over short and long distances, back and forth across the brain and down into the body. We use this activity to create memories and importantly, our thoughts and feelings about memories. Going back to the neighborhood analogy, you could think of electrophysiological activity in neurons as people chatting with one another locally within their own house and sometimes with those who live in houses nearby. However, this activity can travel longer distances using various mechanisms that might resemble picking up a phone and talking or getting online.

PREPARING FOR THE FUTURE

Think about the brain on a minute-by-minute and day-to-day basis. It doesn't know exactly what might happen, but it must be prepared for what could happen. How does it prepare? In the mid-twentieth century, the Canadian psychologist Donald Hebb published his book *Organization of Behavior,* wherein he wrote: "Any two cells or systems of cells that are repeatedly active at the same time will tend to become 'associated' so that activity in one facilitates activity in the other."[2] Colloquially, this idea has been shortened to: "When cells fire together, they wire together." By inference, Hebb suggested that neurons might remain actively linked even after an experience is over. This idea fits the description of a mechanism in the brain known as *long-term potentiation* (LTP), which is a long-lasting increase in what is known as *synaptic efficacy.* This phenomenon can be measured with tiny electrodes that are placed in or around neurons. After a brief stimulation to a neuron or set of neurons, the connections among them become stronger; they are "potentiated." And the increase persists over time, sometimes indefinitely. Returning to the neighborhood analogy, it would be as if all the gravel streets between houses were now miraculously paved and had streetlights illuminating the path.

LTP can be found in many brain regions, including the hippocampus, the part of the brain that we know is involved in memory formation. But we still don't know exactly what, if anything, LTP does for us in our everyday lives. It is plausible that a mechanism like LTP could be used by the brain to enhance the likelihood that information gets encoded when things start getting stressful. Imagine once again that you are walking down a dark street alone in the middle of the night. You are already on guard and you hear a person walking up fast behind you. The brain quickly realizes the potential danger and increases the strength of communication between neurons that are being used to process the incoming information—just in case the situation turns out to be dangerous. If it is, the connections between all the neurons involved in the experience are stronger and will likely remain intact to help re-create the memory later. If, on the other hand, the person behind you ends up being the neighbor, no real loss.

Back in the '90s, neuroscientist Louis Matzel and I suggested a similar idea for LTP—that it might prepare the brain for learning by enhancing the salience of information that is processed during a traumatic or stressful life experience. Specifically, we hypothesized that something like LTP would be induced when we become frightened, and this process heightens the salience of stimuli in our environment so that we can experience and respond to them wisely and more rapidly. It would be as if you increased the signal while reducing the noise from the background. This idea was met with fierce resistance because, at the time, LTP was considered more of an all-purpose memory mechanism, not one made for modifying the strength of stressful memories.[3] But being alert and ready for what is happening now gives us more control over the future, at least to the extent that we can be better prepared to learn when something bad is about to happen. And a mechanism like LTP could come in handy during stressful times, allowing us not only to respond quickly but also to learn and better remember what happened.

Certainly, one might imagine that these kinds of brain processes could be used to help us expand or *potentiate* our view of what is threatening and could cause us harm. But again, studies in rats may be hard to extrapolate to humans. Let's consider a human study conducted by neuroscientists Joseph Dunsmoor and Elizabeth Phelps at NYU.[4] Participants agreed to be exposed to a brief shock before, during, and after viewing pictures of various objects of no real significance in a variety of categories. Later, the memories of the previously neutral objects were more potent if the objects came from the same general category as objects that had been associated with the shock and the feeling of fear. Somehow the brain anticipated what could happen and was ready to retroactively enhance the salience of some stimuli to the exclusion of others. This finding is one among many indicating that the brain has its own way of generalizing the feelings of fear—by expanding from what caused fear to other similar stimuli in the environment.

Generalization is a relatively natural process, which occurs often in the absence of stress or trauma. In fact, even young children quickly learn these kinds of associations. For instance, without any explicit training, they can easily categorize a poodle and a beagle as "dogs," but will not similarly categorize a poodle and a turtle. And for most of us, this process is an adaptive response, at least in the short term. If a lion is chasing me across the savanna, it is good to remember what the animal looks like and avoid animals that look like that in the future. However, in our current society, this process can get out of control and cause us to have undue stress and fear in the present because what we are experiencing now reminds us of what happened in the past. As mentioned, some people who experience traumatic events begin to generalize what happened to situations that are not likely to cause harm and end up behaving in ways that are not adaptive. Paradoxically, the thoughts and the behaviors that were generated by the initial experience then become harmful in and of themselves. For example, recall the husband and wife in the tragic car accident that I discussed at the beginning of the book. Even though the accident was long over and both were now physically safe, the wife was finding it difficult to drive and did not feel comfortable leaving the house. Again, in the short term, some of these responses are useful and always understandable. But with time, they need to be replaced with thoughts and behaviors that are more consistent with the danger that is present in this moment—and the realization that danger may be absent.

THE MAKING OF MEMORIES

Memories capture our interest, for instance, when reminiscing about good times in the past. But few of us have given much thought to how the brain goes about making a memory in the first place. I used to be the same way. Even as an undergraduate, memories did not interest me that much. But as a doctoral student, I had the good fortune to attend a lecture by (and later work with) one of the most famous memory researchers then alive—Dick Thompson. He was the first person to find a memory trace in the mammalian brain.[5] During his lecture, he asked the audience to think about memory—*How in the world does the brain make a memory? How do we learn? How can a piece of tissue made up of protein and fat take what we experience in everyday life and keep a record of it to relive over and over again? How does this even work? Not only that, how does it happen so fast?*

Memories are made nearly instantaneously. Think about that for a second. Do you remember the name of the scientist who gave the lecture? Dick Thompson? You probably do because I just mentioned him. If you don't remember it, you certainly recognize it. And you recognize it now—just seconds later. This is how fast memories are made. Similarly, think about something you did earlier today, which you can now clearly recall. How can the brain make memories so fast?

And how do they last so long? Of course, some of our memories get replaced with more recent memories—such as where we parked the car at the grocery store. But others stay with us forever—especially the traumatic ones. Rutgers neuroscientist Louis Matzel once told me about an experiment of his where a mouse was trained to run down an alleyway over and over again to get a piece of food. Then one day, the mouse ran down the alleyway, and instead of the expected food reward, it received a mild shock to its feet. After that, the mouse was not so eager to run down the alleyway to get the food. But eventually, the mouse would take a chance and run down the alleyway. Then the mouse was put back in its cage for a few weeks, which is a long period of time in the life of a mouse. Later, the mouse was given the opportunity to run down the same alleyway. What did it choose to do? Nothing. It did not even try to get the food. Of the two past experiences—a good one and an aversive one—the aversive one took precedence.

Once again, we can only surmise what a laboratory mouse or rat remembers from nonverbal behavior. But when we are studying people, we can ask them. In one of our studies, we asked a relatively large group of adult women to reflect back on the most stressful day of their lives.[6] We did not ask people to tell us *what* happened, but rather wanted to know *how* they experience the memory in real time now and *how strong* it felt. Some of the women had a history of sexual trauma, while others did not. The women who had this history said that they experience their most stressful life memory as intense and especially vivid. They particularly remembered the time and place where it happened, such as the house, room, furniture, or the time of day or year. Their memories often included the people who were around and what they were doing. In general, the women in our study said that they tended to remember the sequence of events happening across time. It was as if they could "see it" like a movie in their heads from when it started to when it was over. This is consistent with what many people say happens to them after a traumatic event—they feel like they cannot stop playing the scene over and over again. And as expected, the women said that the memory represented a significant event in their lives, one they would not soon forget, and which often dominated their thoughts. They also said that they were ruminating more often than before. Not only that, their tendency to ruminate correlated with the vividness of the memories. And thus they were ruminating more on more vivid memories, potentially making still more of them in the brain.

INDELIBLE IN THE HIPPOCAMPUS

One day in the fall of 2018, I was watching TV when a chyron came running along the bottom of the screen: INDELIBLE IN THE HIPPOCAMPUS. I immediately perked up. Then my phone started ringing off the hook. Reporters were calling to find out what they could about the hippocampus—the part of the brain used for making memories. The United States Congress was in the process of confirming Judge Brett Kavanaugh to the Supreme Court. During his confirmation hearing, Kavanaugh was accused of sexual assault by a woman he met during high school. The woman, Dr. Christine Ford, now a clinical psychologist, was asked during her testimony to talk about what she remembered from that fateful night, some thirty years ago. She said she remembered going to a party and later going up the stairs to the bathroom, when she was pushed from behind into a bedroom. She said she remembered the layout of the room, where the bed and dresser were, and so on. In particular, she said that she remembered the voices of two boys laughing. She said this part of the memory was "indelible in her hippocampus."

My lab had just published a study about sexual trauma and memory, and as a result, the press wanted to know what I knew.[7] Some reporters were interested in the memory, but most wanted to know whether the alleged assault actually happened. Obviously, I do not and literally *cannot* know what happened. But it can be instructive to consider her memory of that day. Dr. Ford said she remembered what happened after she was upstairs and seemed to especially recall the order in which events happened. She said she could *not* remember much about the party beforehand or what happened after she escaped the house. This is common and understandable for someone who has been through a trauma. When she was on her way to the house or hanging out downstairs, her brain couldn't have known what would happen in the future, and nothing particularly bad had happened with these people in the past. Therefore, it wasn't prepared to capture all those memories—those everyday memories that most of us would not necessarily be able to recall. But once she became afraid, the body activated all the systems that I discussed earlier. She would have released epinephrine and cortisol from her adrenal glands. In a short time, these neurochemical processes would have infiltrated her body, eventually priming her brain to encode memories of what was happening. Perhaps she even initiated some long-term potentiation among the connections in her brain. According to her recollection, those memories are still with her today.

People often think that memories are stored in the hippocampus, which is not exactly true. The hippocampus is used for making memories and is especially engaged while making memories about ourselves—the memories that make up the stories of our lives. How do we know this? In certain cases, a person must have their hippocampus surgically removed because of seizures, and afterward they have a hard time making new memories about themselves and their surroundings. They likely would not remember what they had for breakfast that day or why they had an argument at dinner the night before. But importantly, they would tend to remember things they knew before their hippocampus was excised. For example, they would likely recognize their life partner and children and would know how to get around their house.

The hippocampus is arguably the most studied of brain regions. But despite our efforts to understand it, we still do not know exactly all that it does and how it goes about doing it. We do know that it is involved in generating some kinds of memories and in integrating those memories with others. The memories it tends to produce are usually "declarative," meaning they depend on conscious awareness and can be described with words. Sometimes these types of memories are referred to as "episodic" because they encode episodes in our life. And they are most always autobiographical. You might say that the hippocampus helps us tell and retell the stories of our lives.

However, not all memories generated by the hippocampus are dependent on language. All mammals have a hippocampus, which they use for learning even though most do not talk. But like us, other mammals use their hippocampus to encode where they have been and what they were doing while they were there. In a very general way, the hippocampus helps us navigate in space and time. The structure even contains cells that are designed to encode space (and by connection, time). These so-called place cells fire when we move around and then somehow integrate with other neurons and brain regions to generate a type of record of where we have been, at least recently. All this information provides a representation of the context of our present experience that we can use later to predict what might happen in the future and, if necessary, retrace our steps.

If I were to stick an electrode into your hippocampus right now, the neurons within it would likely be firing—a lot—and they would be firing to whatever it is you are doing now, including reading this book. The hippocampus is involved in many activities that matter to us in our everyday lives. It is always working. And much like the brain itself, we would not be who we are without it. We need it at all times, but especially when danger is present.

I like to think of the hippocampus as a real-time learning machine. It takes what is happening now and how we feel about what is happening and associates it with what has happened in the past and how we feel about those past memories, all in the service of making predictions about what could happen in the future. As such, the hippocampus is always online and always learning, encoding this moment in time and space and then integrating it with memories from the past. The hippocampus somehow binds all these memories together in a story that we then tell ourselves—sometimes over and over again. In one study, researchers induced sadness in people with and without a working hippocampus.[8] Those without a working hippocampus were sadder for longer. Perhaps they could not break away from their feelings. Maybe they were trying to figure out why they were sad and couldn't. Perhaps they couldn't remember what made them sad in the first place. In a very general sense, the hippocampus links our past with our future and our thoughts to our memories.

RESURRECTING THE FEELING

Thinking back over past traumas can be bad enough, but what really makes the thoughts unbearable are the feelings that come with them. But the feelings themselves do not come from the brain. The brain does not really have feelings, at least not in the general sense of pain or fear. Indeed, it is quite difficult to feel physical pain once it is gone, even though we can clearly remember that it was bad. The essence of our feelings comes from activities within the peripheral nervous system—mostly the autonomic nervous system that I discussed earlier. So how do feelings even arise? How can a thought dredge up an old memory and then attach that memory to a feeling in the body? One way is through the amygdala, a structure deep in the sides of our brains and close to the hippocampus. The amygdala has connections to the rest of the body and as such can instigate the feelings that we often associate with stress and fear and trauma.[9] I once saw a video of a man with electrodes implanted in both his amygdalae. When the electrodes were activated, he said he did not feel pain, just fear. At one point, he said he was terrified, as if he were about to be bitten by a big dog. When only one amygdala was activated, he said that he still felt fear but only on one side of his body. And when select parts of the amygdala were activated, he said that he felt like he could control the fear more than when other parts were activated.[10] Now, of course, this is not a normal experiment, and a true and complete feeling of fear can not be elicited by stimulating an electrode in someone's amygdala. But these observations suggest that some aspects of our everyday experience, including those that generate feelings of fear, are governed by this structure and its connections to other structures in the brain and throughout the body.

In some cases, the amygdala seems to be able to enhance the memory for fear even after the fact. Take the studies done by Jim McGaugh at UC Irvine back in the '60s.[11] Rats were exposed to a stressful event and then tested to see if they remembered the event later in time. Most of them remembered it very well. But if activity in the amygdala was disrupted right after the stressor, they didn't respond to a reminder, as if they did not remember what happened. When McGaugh and his team bypassed the peripheral nervous system entirely and simply injected adrenal hormones into the amygdala, the memory for the event was strengthened. In fact, just stimulating the amygdala was sufficient.[12] This system seems to work similarly in humans, at least under far less stressful conditions. People were in surgery for epilepsy and agreed to have their amygdalae stimulated while the experimenters presented them with various pictures to remember. The participants remembered the items better when their amygdalae were stimulated right after they were presented with the images.[13] In real life, we presume the amygdala does this for us when we need it most—to enhance the memories of those unexpected and yet significant moments in time. But we also need to learn how to turn it off, so to speak. As someone once told me, "My amygdala was on autopilot during the trauma, and I am still trying to turn it off."

THE MANY TRACES OF TRAUMA

The hippocampus and amygdala are constantly communicating with each other and the rest of the body to record and then reconstruct a comprehensive memory for events in our lives. Let me illustrate this process with a study by Hanna and Antonio Damasio, who lead the Brain and Creativity Institute at the University of Southern California.[14] The research team had access to three people: the first person did not have a working amygdala, the second person did not have a working hippocampus, and the third person had neither of these brain regions functioning in a normal way. The people agreed to have a mild shock delivered to one of their arms. Right before the shock, a tone was played. Someone with a fully functioning brain would quickly learn that the sound of the tone meant the shock was going to happen and would flinch when they heard the sound, and they would start to sweat. Meanwhile, they would also become consciously aware that the tone predicts the shock and would be able to say so. But what about the people who did not have a working amygdala or hippocampus? The researchers observed that the person without a working amygdala didn't sweat when presented with the tone. But he was consciously aware of what the tone predicted. He could tell someone in words that the tone meant that the shock was coming.

In contrast, the person without a working hippocampus could not tell someone (or himself, for that matter) that the tone predicted the shock, even though his body still responded to it—he sweated. Finally, the person without either brain region working could not verbalize the meaning of the tone, and neither did he sweat. These results are consistent with the idea that the hippocampus is used for conscious reflection on fearful experiences—knowing what and where something happened or when it is about to happen. Meanwhile, the amygdala is more involved in the body's response to the memory of it—the fear response. But even more importantly, these case studies suggest that the brain is full of trauma memories, different kinds of them for generating different kinds of thoughts and responses throughout our bodies, and for use at a later time.

To be clear, fear is not located in the amygdala any more than memories are located in the hippocampus. If I could take out my amygdala and lay it on the table, would it feel fear? If I could take out my hippo-campus and lay it on the table, could it produce a memory? Of course not. It is just a piece of tissue. Through electrical current, these structures interact with the rest of the brain and the peripheral nervous system to help us learn and remember. When something bad happens, the fast fear response alerts the brain to encode what is happening and to make quick decisions. Meanwhile, the slower stress response is releasing cor-tisol into the blood, which eventually reaches the brain to modify and in some cases, create the structures we need to make a lasting memory.

A BRAIN WANDERING IN THE PAST

People who have been traumatized often think about what happened, going over autobiographical memories, revisiting what happened or imagining what could have happened. These memories are often ac-companied by ruminative thoughts about themselves and the feelings they have about themselves.[15] These kinds of thoughts not only dis-tract us from what is happening now, they also change our brains. But how do they change our brains? As you can imagine, because these thoughts are wide-ranging and complex, most regions, and indeed most neurons, probably get involved at some point along the way. But there are combinations of regions—called *networks*—that seem to work together while we ruminate. One study analyzed imaging data from nearly three hundred people who had participated in var-ious studies on rumination.[16] The researchers focused on a network referred to as the *default mode network*—or DMN. This network comprises groupings of brain regions that are mostly within the out-side layers of the brain and are implicated in PTSD.[17] These regions are engaged together when we let our minds wander around with no goal or focus of attention. In general, ruminative thinking seemed to engage this network more than others, and some brain regions were especially active, such as those within the prefrontal cortex, or front

of the brain. This finding is not especially surprising because ruminations are autobiographical in nature, and we need to engage the
prefrontal cortex as we ponder, wondering how we feel about what
we have been doing with our lives.

However, another of their observations was more surprising and
potentially insightful. Ruminative thinking was associated with *less* activity in networks within the temporal cortex, the area around the ears,
which is involved in learning and memory and only a synaptic connection or two away from the hippocampus. Indeed, these temporal
regions are especially active while people are learning to differentiate
(i.e., discriminate) between similar patterns of information, such as
memories of what happened in the past from the experience of what is
happening now.[18] Why would these regions be quiet as we ruminate?
Perhaps when we are caught up in repetitive thought patterns, we aren't
engaging parts of our brains that we use for new learning, including
the hippocampus and adjoining regions. Maybe we can't engage them,
even though we could most certainly use them to encode what is happening now and distinguish it from what has happened in the past.
Sure, these kinds of comparisons come in handy when we are trying to
make plans for dinner and don't want to eat the same thing twice, but
we also need them to make more vital decisions about how we feel and
what we should decide to do in this moment. These kinds of comparisons help us know when we should generalize and when we should not.
They help us know when we are safe. Maybe the brain is simply not
as flexible when it is too busy ruminating on what was or could have
been. It's almost as if our brains are stuck in the past, using the same
networks over and over again as we tell the same story to ourselves.

SLOWING DOWN TO WATCH

You might have seen the movie *Eternal Sunshine of the Spotless
Mind*. The main character, played by Jim Carrey, is heartbroken

when his girlfriend breaks up with him. He can't stop thinking about her and their love affair, and so he decides to have a brain procedure to erase the memories of their times together. But in the process, he loses the good memories as well. This was not what he wanted; he wanted to remember the good times while forgetting the bad. Now, in reality, neuroscientists have not yet figured out how to erase a memory, much less erase the bad ones while leaving the good memories intact. These memories reside in our brains, and even though we may not like all of them, they aren't going anywhere for now.

Since it appears that we can't get rid of the memories of trauma, perhaps we can just stop thinking about them. That was the opinion of one Mildred Norman. Otherwise known as Peace Pilgrim, Mildred Norman was born in rural New Jersey in 1908. She spent her early adult years as a secretary and sometime flapper near Atlantic City. She was married briefly, quickly realizing that marriage and housework were not for her. In fact, home was not for her. In her forties, she walked the entire length of the Appalachian Trail and soon thereafter gave up all her possessions, crisscrossing the United States seven times on foot. She walked for nearly thirty years, logging over twenty-five thousand miles before she quit counting. She obviously had a lot of time to think, all of which culminated in this: "If you realized how powerful your thoughts are, you would never think a negative thought."[19]

It sounds good—to never have a negative thought. But most of us don't have thirty years to figure out how to do it, if it is even possible. But perhaps with insight into the process, we can redirect some of our thoughts before they bring up those bad memories or maybe we can at least watch them as they play themselves out, without letting them connect to so many other thoughts and feelings. Most of the time, we start thinking about something, and before we know it, we are connecting that thought with memories. Some of these memories are bound to be negative, and then they bring up feelings from down in the body. For most of us, this process—going back and forth from thoughts to memories to feelings and back again—feels instantaneous. But it is not. It takes time for information to travel around the brain and the rest of the body. It takes time to re-create the story. Therefore, with time and practice and some insight into how it works, we can learn to slow down and examine this process as it is playing out.

I remember a time long ago when slowing down and watching this process play out in my own brain really helped me out. I had prepared dinner for my son, and he acted like he didn't want to eat it. I am not a big cook, and I could feel myself starting to get angry—I could see the memory of myself cooking the meal and could feel my blood pressure rising. I could almost feel my mouth getting ready to blurt out something mean. But instead, I quickly went upstairs and sat with my thoughts and feelings and let them dissipate. It turned out that he was sick—physically ill—which was why he did not want to eat. Luckily, I had slowed down long enough to see where my thoughts were leading me before I said something hurtful to my son. With these kinds of skills etched in our brains, we can learn to make finer distinctions between old and new thoughts, between those that are adaptive versus not, and between situations that are dangerous or not. We can decide for ourselves: Do we really want to linger with the thought long enough for it to dredge up that old traumatic story? Do we really want to give it time to connect with all those feelings in the body? Maybe we can even learn to redirect our thoughts and memories for a better quality of life, but more on that in part 3.

6

Women and Their Changing Brains

Upon those who step into the same rivers, different and ever different waters flow down.

—Heraclitus of Ephesus,
Greek philosopher (540–475 B.C.)

The facts are indisputable: women are more likely to be diagnosed with PTSD and depression; they are more likely to be diagnosed with anxiety-related disorders and social phobias; they are vastly more likely to be diagnosed with eating disorders such as anorexia and bulimia. Why is this? Is there something about a woman's brain that makes her more vulnerable to these stress-related conditions? Or is it all because of our upbringing and the differing ways we experience life in modern society? I've been asking myself these questions for nearly thirty years now and have concluded that there are several reasons. Let's start with why it took us so long to find answers to these questions, or to even ask them in the first place.

WHY HAVE WE HEARD SO LITTLE
ABOUT WOMEN AND PTSD?

Most of the early research on trauma, especially PTSD, was conducted with men and, in particular, men who came home from war.
During the Civil War, soldiers often expressed a distressing constellation of symptoms. Physicians, who believed that the men were
homesick and nostalgic for a more peaceful past, referred to their
condition as *nostalgia*. During World War II, similar symptoms were
noticed in military personnel, but in this case, they were said to be
shell-shocked. Physicians thought soldiers' brains were damaged by
shells exploding from guns. It wasn't until the Vietnam War that
the term *post-traumatic stress* and its corresponding disorder, PTSD,
were introduced. Many Vietnam soldiers came home to the United
States complaining of problems with anxiety and intrusive memories
of what had happened during combat. They also came home with
other problems, including alcohol and drug abuse. Thanks to veteran groups and government funding, the soldiers finally began to get
some attention and the help they sorely needed. Of course, women
served in these wars as well, but they were less likely to engage in
combat, which we know is more likely to induce PTSD. Moreover,
the number of men enlisted compared to the number of women was
not even close to equal. As a result, the disorder became associated
with men who have gone to war.

To this day, people still associate PTSD with men, and until quite
recently, most research on this topic was conducted almost exclu-
sively in males. Clinical studies were often carried out at local veteran
hospitals and associated institutions, which primarily treat men. And
clinical trials often refrain from enrolling women because they could
be or could become pregnant. And, in laboratory studies, which tend
to test mice and rats, male animals were nearly always used. The reasoning behind this practice did not have much to do with PTSD
or which sex was most vulnerable to trauma. Rather, the reasoning

was related to sex hormones. Female mice and rats go through what is known as an *estrous* cycle, which is similar to a menstrual cycle in women, but occurs about every five days. During both the estrous and the menstrual cycles, estrogens and progesterone are released from the ovaries into the blood. Then the presence of these hormones decreases and increases again for another round of ovulation, assuming fertilization and pregnancy have not occurred. These changes in hormones can influence behavior. For example, female rats in *proestrus,* when estrogen levels are elevated, learn some laboratory tasks better than female rats in estrus, when estrogen levels are low.[1] On the basis of these kinds of data, it was widely believed that introducing females, on different cycle days, into experiments would cause more variability—more noise—in the data. Minimally, it would require more experiments and more funding.

Oddly enough, when neuroscientists wanted to know what estrogens did to the brain, they continued to test males. I still remember hearing a scientific lecture back in the late 1980s where the scientist talked about injecting rats with estrogen and measuring electrical activity from neurons in the hippocampus. During the question-and-answer period, I asked the speaker whether the female rats still had ovaries. He replied that he tested only males. I was stunned. As ludicrous as it may sound, it was generally assumed that the female brain was just a male brain with some estrogen thrown in. But luckily, those days are over. Now we have a mandate from the National Institutes of Health, which funds most of this research in the United States,[2] to include both sexes, unless the researchers have a very good reason not to do so. This was a significant achievement in the field and one that I was not sure would happen in my lifetime.

To be fair, there is another reason neuroscientists did not study both sexes for so long. Most of us assumed that basic mental processes such as learning and memory, or even how the brain responds to stress and trauma, would be more or less the same in males and females. Most sex differences in the brain were thought to involve sexual behaviors—the act of sex and its role in reproduction. But now we know better. Let me tell you about a series of laboratory studies my lab did with neuroscientist Debra Bangasser that illustrates just how wrong we were.[3] Male and female rats were exposed to a stressful event and then trained to learn something new the next day. The males learned better while the females did not learn well at all. However, if the female rats were given testosterone when they were born, they would grow up to act like male rats, meaning they now learned better after stress. Then we started inactivating different parts of the brain to see if males and females were using the same or different brain regions to respond to stress, and we found that males were using different parts. In particular, they were using a part of the brain known as the *bed nucleus of the stria terminalis,* or the BNST. However, when the females were given testosterone at birth, they grew up and used the same brain regions as the males. These results in rats are fascinating, but whether they inform us about sex differences in PTSD in humans is debatable. Minimally, they indicate that males and females can use different brain regions to learn from stressful life experience. They certainly indicate that we were wrong to think that sex differences in the brain only matter for sexual behaviors and reproduction.

SEEKING HELP

Now that we have overcome some of our biases against studying females in the laboratory and have accepted the fact that women are more likely to be diagnosed with PTSD and other stress-related disorders, what can we do about it? First of all, we have to come to some agreement as to why women are diagnosed more often than men. This is not so easy because the reasons are complex, complicated, and controversial. First, we must consider the process of diagnosis itself. It is often assumed that women are more likely to be diagnosed because they are more likely to seek help. And there is some truth in this. It is certainly true that many men, even those who are severely trau-matized, are reluctant to seek professional help. Growing up, I had an uncle who served in World War II. He was one of several soldiers occupying a tank that rolled over his best friend. My uncle survived, but his best friend did not. I remember hearing my relatives talk about how this experience traumatized my uncle. Apparently, he never dis-cussed the event, let alone his feelings. It was taken for granted that because he was a man of a certain era, he would stay silent.

Women, on the other hand, were and still are expected to talk more freely about their feelings and show more emotion. Even the word *hysteria,* which refers to uncontrollable emotion, comes from the Greek word for *uterus.* Some physicians actually believed that a woman's uterus moved around in her body, causing her to emote. Thankfully, this word is no longer accepted in professional circles. And more important, some of the stigma associated with mental ill-ness has lessened. All that being said, when my research group re-cruits participants for studies about trauma, women are much more likely to volunteer. Frankly, I see this as a positive response. They are seeking help, and help is available.

THE KIND OF TRAUMA ITSELF

There is another factor to consider—the characteristics of the trau-matic event itself. As discussed, men are more likely to go to war and therefore more likely to suffer PTSD from combat trauma. In contrast, women are more likely to experience sexual violence and therefore are more likely to suffer PTSD from this kind of trauma. Indeed, of all the traumas someone can experience, sexual violence is most likely to induce PTSD.[4] It is estimated that one in three women experience sexual or physical violence in their lifetimes,[5] and the numbers add up—and up. These experiences tend to happen when women are young and particularly vulnerable. And gender differences arise even during childhood, with boys more likely to be exposed to domestic violence and girls more likely to experience sexual abuse.[6] Now obviously, men who experience sexual trauma are also vulnerable to the symptoms of PTSD, but fewer men have and report these experiences than women. As a result, many women are diagnosed with PTSD because they are more likely to experience and report sexual and physical violence, whereas more men are likely to be diagnosed because they experience combat and other types of trauma, with some less likely to seek support.

So, the high number of women diagnosed with PTSD can be attributed, in part, to the types of trauma that women tend to encounter and the disturbing prevalence of interpersonal violence. But this factor does not explain everything. Take, for example, the massive earthquake that hit the Italian city of L'Aquila during 2009. The earthquake itself lasted only twenty seconds, but more than three hundred people died, and thousands were left traumatized. One study examined symptoms of PTSD in a group of people who were directly affected by the earthquake and compared them to symptoms in another group of people who lived close by and were unaffected. Women who were directly exposed to the earthquake endured more trauma symptoms than men did, even though they were all exposed to the same traumatic event. The women said they tried to avoid reminders of what happened but remained troubled by the memories, which they rehearsed over and over again in their minds. Men who were in the earthquake, on the other hand, reported that they were not as affected by the memories themselves yet were more likely to engage in reckless behaviors such as substance abuse.[7] Another study compared responses of men and women who were present during a bank robbery, finding that women experienced more symptoms of PTSD, as well as feelings of fear, horror, and helplessness.[8] Studies such as these suggest that the specific details surrounding the traumatic event, albeit important, do not fully explain why women are more vulnerable to stress-related mental conditions.

THE POWER OF SEX HORMONES

There is more than enough evidence to support the hypothesis that sex hormones contribute to sex differences in PTSD, and other stress-related mental illnesses such as depression. For one, sex differences do not tend to emerge until puberty.[9] In other words, the numbers of boys and girls diagnosed with stress-related mental illness are similar between the sexes in childhood, but as females begin to menstruate and become sexually active, sex differences emerge. Moreover, some women experience mood changes across their menstrual cycle, again as estrogen and progesterone levels are fluctuating up and down. For those who give birth, the act of childbirth itself is usually quite stressful (albeit joyful) and in some cases traumatic. After giving birth, some women go on to experience a form of depression known as post-partum depression and in rare instances, they can even experience a form of psychosis. Then, as we age, some of us experience changes in mood during perimenopause and menopause, as estrogen levels diminish. So, sex differences in the diagnosis of stress-related mental illness, such as PTSD and depression, are mediated in part by changing concentrations of hormones as women go through their everyday lives.

And these changing hormones change the brain. Once released into the blood, they enter the brain, where they bind to receptors, in the same way that cortisol binds to receptors. And then, like cortisol, they go on to change the structure of the brain. Previously, I described tiny anatomical structures called spines, which are used to make connections between neurons. In one amazing study, rats injected with the sex hormone estrogen produced as many as 30 percent more of these spines in the hippocampus. Moreover, these changes occurred naturally.[10] As mentioned, female rats ovulate every five days, and as estrogen levels increased, so did the number of spines. As estrogen levels went down, so did the number of spines. There is no reason to think that a similar process does not happen in women. For example, one study reported anatomical changes in the hippocampus and temporal cortex as a woman went through her menstrual cycle.[11]

Obviously, sex hormones are powerful. But importantly, they do not completely "explain" gender or even sex differences in mental health. They are still just hormones. We probably shouldn't even call them "sex" hormones, because females can produce testosterone, and males can likewise produce estrogens. Neuroscientist Robert Sapolsky once said something to the effect that "the influence of our thoughts and behaviors on our hormones is probably stronger than the influence of our hormones on our thoughts and behaviors."[12] I agree and would take his comment a step further: our thoughts and memories change the brain, which then in turn changes our thoughts and memories, which then change the brain, and so on. We are living within a series of feedback systems that are always circling back and forth and around. They help us respond to stress and trauma in the moment, and then they help us recover. They help us go through the various stages of life with memories of what happened etched in our brains so that we can prepare ourselves for what will happen in the future, good and bad.

MOTHERHOOD AND BEYOND

Actress Sophia Loren is credited with saying, "When you are a mother, you are never really alone in your thoughts. A mother has to think twice, once for herself and once for her child."[13] Before giving birth, many of us go through life focused primarily on ourselves. If we become a mother or caretaker, our attention shifts rapidly. We are now in charge of keeping this helpless little being healthy and alive. All we can do is try to stay focused and learn as fast as we can. Most mammals begin having sex around puberty, and most females eventually become pregnant. After they give birth, they are different, and their brains are most certainly different. In particular, the maternal brain responds to stress differently from a nonmaternal brain. While a graduate student in my lab, Benedetta Leuner examined how the maternal brain responds to stress. We had already reported that a virgin female rat did not learn well after stress; just the presence of an adult male was sufficient to impair learning.[14] But if the female rat had given birth and was taking care of her offspring, she learned well, even with the aggressive male rat lurking around. If anything, she was especially alert and ready to protect her offspring. These data suggest that something had changed in the female brain so that now as a mother she was not adversely affected by the stressor. You might even say that the new mother had become resilient. This change in her behavior was not due to hormones or even the act of giving birth, because foster mothers were also immune from stress. Virgin female rats will, under the right circumstances, learn how to take care of the offspring from another rat. Obviously, they can't feed them, but they will gather them together and sit on top of them to keep the young ones warm and away from danger. In this study, virgin females that were exposed to a stressful experience could learn well, as long as they had learned to take care of the young rats, even if they were not their genetic offspring. And interestingly enough, this effect seems to persist. Mother rats remained resilient long after the young rats had left the nest, so to speak.[15] As an empty nester myself, I take solace in these data!

Recently, I heard a podcast with Australia's Bronwyn Graham, who is studying how mothers learn about fear.[16] Mothers and women who were not mothers were shown the face of a man and soon thereafter felt a brief shock to their hand. Then they were shown the face of another man but given no shock. Once they learned that one face predicted the shock and the other face did not, both faces were presented without the shock, again and again—a process known as *extinction training*. The researchers measured fear by recording how much the participants sweated when presented with the faces. The mothers learned especially well during extinction training, meaning they expressed less fear in response to the face when it was not associated with the shock anymore. These results support the general notion that mothers adapt quickly to changing conditions and contingencies.

Obviously, these are artificial conditions created for an experiment; therefore, we don't really know the value of these responses. And so, to be clear, I am not claiming that mothers are necessarily better or worse at learning than women who are not mothers. Nor am I claiming that they are better or worse at responding to trauma. But studies and everyday observations do suggest that their brains are primed to pay particular attention to select cues in their environments, which are important for maintaining the health and well-being of their offspring. Moreover, studies indicate that the actual structure of the brain changes after giving birth and remains this way.[17] It is not hard to imagine why changes in learning processes would emerge. Mother brains need to be flexible.

INCLINED TO RUMINATE

For more than ten years now, I have been asking everyday people—mostly women—about their thoughts. When I tell them about ruminative thoughts, they emphatically nod, as if they can completely relate to what I am saying. In fact, I was recently at an event, speaking about ruminations. After my talk, a woman came up to me to tell me that she had just been offered a job that she really wanted. She had written her new boss an email to tell him how excited she was about the offer. After she sent it, she found a misspelling—a typo. "Now," she told me, "I can't stop thinking about the typo and worrying about the impression it made on my new boss." Why was this woman so focused on this one thought? Why had it taken up so much space in her brain? Why did she keep blaming herself for an innocent mistake that her potential boss probably didn't even notice?

Numerous studies, including some of my own, indicate that women ruminate more than men do.[18] And many of those ruminations are about far more serious events than typos in an email. Exactly why women ruminate more is more difficult to answer. There are several theories. One suggests that women ruminate more because they tend to focus on their internal feelings more—*Why am I feeling so sad? Why am I so nervous?* As a result, they are more familiar with their own thoughts, which makes those thoughts easier to rehearse over and over again. Another theory poses that because women are more likely to suffer sexual and physical violence, it can exacerbate the production of these kinds of thought processes.[19] But of all theories, the most accepted is about depression. Women are nearly twice as likely as men to experience depression in their lifetimes, and ruminations are tightly linked to depression.[20] In one of our studies, women who were depressed were likely to ruminate, reflecting on the past. Perhaps they were trying to understand what had happened to them along the way to make them feel the way they do.[21] But to be clear, men do ruminate, and they similarly tend to ruminate more when they are depressed.

If ruminations really do contribute to the high incidence of stress-related problems in women, then it would be good to know why women tend to ruminate. Perhaps it has to do with control. Indeed, one study reported that women who tended to ruminate said they did not feel like they had much control over their lives. They also said that they used ruminative thoughts to deal with their feelings.[22] Let's consider a concrete example. Imagine I am in my late thirties and trying to meet the man of my dreams. I want to get married and have children, and I am getting older. Moreover, I am lonely. I go online and venture out on date after date, but each date is worse than the last. Some are even traumatic. After a number of bad experiences, I just give up on dating and men in general. I feel helpless to get what I want. I've learned that by trying to help my situation, I only made myself lonelier and more despondent. I've learned *not* to respond. And I can't quit thinking about what happened—or didn't happen.

But we can also learn how to gain a sense of control. Imagine instead that I go out on a super fun date the very first time out. Even though it does not work out in the long term and he is not the man of my dreams, I am hopeful that I can meet someone through online dating. Now I am less lonely because I have engaged in an activity that could get me what I want, as long as I keep trying. This may seem like a benign example, but it isn't. People want to connect with others. We want to find someone to love. But along the way, we might also learn to avoid others because we become anxious or don't want to be hurt by them. In short, we learn about control by our actions in the world. We can learn to have more control of our thoughts, regardless of our age or sex or gender. But to do so, we must accept the fact that the brain does not control us. It *is* us.

NOT ONE REASON

The reasons why women are more vulnerable to stress and trauma and more likely to be diagnosed with PTSD are many, and not limited to those I have discussed here.[23] When I first started thinking about this problem some thirty years ago now, I was naive, actually mixing up the words *sex* and *gender* (*sex* is used for all species, whereas *gender* refers only to humans). Then I went through a stage where I assumed that most sex differences in mental health could be explained by sex differences in the brain. Now, I am neither naive nor certain, at least in general. But I am fairly certain about a few things: women are always changing, our brains are always changing, and the world around us is always changing. There is good news in this. It means that we can learn new mental skills that help us respond well and wisely in the moment. And once learned, we can take those skills along with us from one stage of life into the next. In particular, I think it would be especially good if women (and men) could find a way to ruminate less often. As I mentioned in a previous chapter, ruminative thoughts seem to be tightly linked to overall mental health, including feelings of depression and anxiety and stress. They can even be linked to the vividness of trauma memories and how someone feels about their body. If we can find a way to lessen these kinds of thoughts, maybe these other thoughts and feelings will fall gently into place.

7

Everyday Neurons for Everyday Life

Knowledge about life is one thing; effective occupation of a
place in life, with its dynamic currents passing through your
being, is another.

—William James, American psychologist

My dad was an accomplished engineer at Standard Oil for nearly forty
years. He loved his job and never missed a day of work and hardly
ever took a vacation. The year he retired, I came home from college
for the holidays to find him sitting in his big chair with his head in
his hands, all day. At one point, my mom took me into the bathroom
and whispered to me that she was afraid to leave him alone. My father was a serious man, but I never thought of him as depressed and
certainly had never seen him behave in this way. We made it through
the holidays, and soon thereafter I decided to move to Los Angeles.
I asked my dad to come with me on the ride out. As we were driving
through the cornfields of Nebraska, he became quieter than usual. I
thought he might be crying, but I was too afraid to look—or ask. At
some point, he began to talk.

My dad grew up in Beemer, Nebraska, a small town with fewer
than a thousand people. His father was the town mortician and
owned the local hardware store and mortuary—a common business

in those days. My dad worked in the store on and off throughout his childhood. One day after working in the store all day, my dad and his brother arrived home to find the garage door shut, which it never was. When they opened it up, they found their father dead. He had taken a hose from the hardware store and hooked it up to the exhaust pipe from the hearse. Before our cross-country trip, I had never really heard the story, certainly not all the details. In my family, like many families, taking your own life was considered a sin, and we did not talk about it. But suicide is everywhere. When I ask my undergraduate class of two hundred students how many of them have been touched by suicide, nearly half raise their hands. And they are only about twenty years old. Suicide is unimaginably traumatic, and for those left behind, the trauma never goes away. We are always left wondering: Why did they do it? What was so bad? Could we have done something to prevent it? I realize now that my dad must have thought about what happened to his own father nearly every day of his life. It was his everyday trauma.

Of all the traumas we experience in life, the everyday ones worry me the most because the damage they inflict on the brain can be the most difficult to turn around. Let me tell you about one example in particular. Neuroscientist Maura Boldrini and her colleagues at Columbia University examined brains from people who had died while suffering with depression.[1] Sadly, a number of them had taken their own lives. The researchers were interested in looking at the hippocampus because we know how sensitive it is to stress. They focused on a part of the hippocampus known as the *dentate gyrus*. This region is often called the *gateway* because it brings information from all over the brain into the hippocampus. And the hippocampus needs all kinds of information to do what it does—make memories of our lives. The researchers wanted to know whether living and dying with depression changed the numbers of neurons in this part of the brain. The neurons in this region are roundish and known as *granule cells*. Indeed, the people who died with depression possessed fewer granule cells compared to people who were not depressed, especially if they had died by suicide. And importantly, the longer the duration of depression, the fewer cells there were. Left untreated, it would appear that depression can have long-lasting effects on the anatomical structure of the brain. And the longer the suffering, the greater the loss.

A SERENDIPITOUS DISCOVERY

When I first started studying the brain in the 1980s, neuroscientists thought of it as relatively stable. We knew it could and did change, but we did not think it could change significantly. Cut ahead a decade or two. I was still studying the brain and still trying to figure out how it makes trauma memories. Imagine my surprise when another young neuroscientist told me that she had discovered new neurons in the adult brain. This might not sound that surprising to you— depending on your age and how much you know about the brain. But at the time, neuroscientists did not think that the brain made new neurons after birth. Indeed, we were taught that it absolutely did *not*. New neurons were not even on the radar!

This process of making neurons is called *neurogenesis,* and the scientist who told me about it is Dr. Elizabeth Gould. She had made a major discovery, one that would fundamentally change the way we think about the brain for generations to come. How had she discovered these new adult-generated neurons in the first place? Like many of us, Dr. Gould was interested in stress and how stress hormones change the brain. In 1989, we were all surprised to read a *Science* study reporting that neurons died when the adrenal glands were removed.[2] Recall that adrenal glands are tiny structures that sit on top of our kidneys, producing stress hormones—cortisol in people and corticosterone in rodents. The dying cells were located in the dentate gyrus of the hippocampus, the same brain region that is so vulnerable to stress and depression. Many scientists set out to replicate this finding, including Dr. Gould. Indeed, she found dying cells in rats without adrenal glands. But she also noticed something even more unexpected. Some of the cells were dividing. Could neurons be dividing in the hippocampus? Was that even possible in adulthood? That is certainly not what we were taught in school. So, she set off for the library. I realize this might sound rather archaic because no one actually goes to the library to read journal articles anymore. But in those days, we had no choice. Lo and behold, she found several papers from the '60s and '70s that suggested that new neurons might be produced in the adult brain—specifically in the dentate gyrus of the hippocampus.[3] Dr. Gould and her group went back to the lab to see if they could find these "new" neurons. Obviously, they did.[4]

USE IT OR LOSE IT

When Dr. Gould asked me if I wanted to see these new neurons for myself, I jumped at the chance. Sure enough, there they were, under the microscope. They actually looked a bit like baked potatoes, all lined up in a row, thousands upon thousands of them. Seeing the cells was exciting for sure. And it raised a number of questions, the most pressing being: What in the world were they doing in there? Dr. Gould's lab and mine decided to join forces and find out. We thought that the new cells might be used for learning. After all, they were being produced in the hippocampus, which is the part of the brain that is used for learning. And we did find that these new neurons are involved in learning, especially if the process of learning requires the hippocampus.[5]

But oddly enough, many of the new cells die within weeks of being generated. Which raises another interesting question: Why would the brain go to all the trouble of making new neurons if many of them don't even survive? And we got to thinking. Perhaps the neurons were being produced for one particular reason, and if they were used for that reason, they would survive. So, we hypothesized that the new neurons were being used to create new memories—but only during a short window of time, and if that time passed, the cells would just go ahead and die. This is perhaps analogous to what happens when you use muscles in your body. You can build up muscles by working them out hard day after day. But if you quit using them and sit around not moving, they disappear. Use it or lose it.

To test our hypothesis, rats were injected with a chemical that labels the new cells as they are being born. Then a week later, just before the cells would begin to die, rats were trained on a task that is difficult to learn. Afterward, trained rats had more new neurons than those that were not trained.[6] Moreover, rats that actually learned had more new neurons than those that were trained but did not learn well.[7] How cool. Learning rescued the young neurons from death. And these changes seemed to persist. Animals that learned still had those neurons months later, which is a long time in the life of a rat.[8] Apparently, once rescued, these neurons stick around.

Finding new neurons in the brain was a big deal, and finding out that learning can keep them alive was also a big deal. And many other exciting findings were still to come. Take, for example, the issue of control. Out in the wild, rats live underground in winding burrows. When confined to close quarters, they establish dominance hierarchies during which one or two male rats start to dominate the other rats. The dominant rat gets easy access to food and water and females. The other adult males become submissive, and for them everyday life is quite stressful. Now, obviously, we cannot ask them how they feel. And besides, rat brains cannot be studied out in the wild. So, Dr. Gould and her graduate student at the time constructed a burrow in the lab. Their burrow had different levels with tunnels and secret places to find food and water. It was then outfitted with infrared lights and cameras to watch what happened at night. Sure enough, one male would start to take control and dominate the other rats. When the researchers counted the numbers of new neurons in the hippocampus, they found more in the dominant male than in his subordinate counterparts.[9] Learning to control the environment enhanced the structure of the brain—by keeping those precious new neurons around.

HARD WORK PAYS OFF

Neuroscientist Louis Matzel once told me that "we are always learning."[10] At the time, I didn't get it. I had tended to think of learning as something we do in school or as a subject to study in the lab, but I have come to appreciate what he meant. We are always learning. In fact, learning is why we have memories in the first place. Memories do not exist to make us miserable or to help us reminisce about our recent holiday vacation. Rather, we need them to learn from our experiences, especially bad experiences. And so yes, we are always learning—always. And importantly, the amount of effort we use when we are learning is always changing. When we learn something new, it is often hard and requires a great deal of effort. But after we learn it, we don't need to concentrate as much because now we know more about it. This is how the learning process works and how one learning opportunity leads to another.

To keep learning effortful, we have to keep challenging and exposing ourselves to new and more difficult learning opportunities. Take medical school, for example. The training is rigorous with lots of memorization and pressure to perform, with minimal sleep or time to rest. But in the end, I would prefer to put my health in the hands of a doctor who has gone through the process. They have learned what they need to do under difficult conditions. In a laboratory study, this type of learning is distinguished from an easier type of learning by the actual time it takes to learn—or the number of trials. If a task requires more trials of training to learn, it is by definition more difficult. I would call it *effortful.* In our studies, tasks that took more trials to learn were more likely to keep new neurons alive.[11] But importantly, learning had to occur. Going back to the medical school example, I would only want to see a doctor who actually passed their exams after all their effortful learning while training in medicine!

GYM RATS ARE SMARTER THAN YOU THINK

Back in the '90s, we were working as fast as we could to find out more about these new neurons, and other labs were doing the same. One group from San Diego noticed that physical exercise could increase neurogenesis. In their studies, mice were given the opportunity to run on a wheel—like the ones you placed in your hamster cage when you were a kid. Mice are not like humans when it comes to exercise. They love to run. They will run miles and miles a day if given the chance. Someone once put a running wheel in the woods. All kinds of animals, mostly rodents, tried to get in the wheel to run—and run they did! Back in the lab, mice were also running in their wheels, and after several weeks of running, the mice had many more new neurons in their hippocampus.[12] When this report came out, labs all around the world stopped what they were doing to study running. Scientists love to study exercise—even if they don't do it themselves!

It turns out that not all kinds of exercise enhance neurogenesis. The exercise needs to be effortful, and by effortful I mean aerobic. The word *aerobic* means "requiring oxygen," and therefore aerobic exercise requires oxygen. When the heart beats fast, more oxygen gets into the blood, which then gets into the muscles—and the brain. In contrast, *anaerobic* exercise does not require oxygen. This type of exercise in humans would include weight lifting or walking. In one study that I especially like, groups of rats exercised aerobically by running on either a treadmill or a running wheel, whereas another group engaged in what might be considered resistance training. Still another group engaged in high-intensity interval training, often referred to as HIIT. At the end of the training program, the group that engaged in sustained aerobic exercise produced more of the new neurons when compared to the group that did resistance training or HIIT. This is not to say that other types of exercise were not of consequence. Overall, the group that engaged in resistance training was physically stronger, and the group that engaged in interval training did have more of the new neurons but not as many as after sustained aerobic exercise. In the end, aerobic exercise over time was the best for creating new neurons in the hippocampus.

What about humans? Does aerobic exercise produce more neurons in their hippocampus? Well, this is difficult to verify. Currently, scientists do not have techniques that can identify individual new neurons in a person while they are alive. Therefore, the studies that do exist are conducted after death in postmortem tissue—like the depression study I talked about—or the evidence is derived indirectly. For example, one study recruited people who do not normally exercise to run on a treadmill or an elliptical machine four times a week for twelve weeks.[13] Others did not exercise. Afterward, the people who had been exercising had more blood flowing into their hippocampus—into the dentate gyrus, which is where the new neurons are produced. Neurons, like all cells in our bodies, are surrounded by tiny blood vessels that bring in oxygen and other nutrients. Without blood, cells cannot divide—or really even survive. We can't know for sure whether the people who had been exercising actually made new neurons—but given the increased blood flow into the region, it seems likely.

DO HUMANS MAKE NEW NEURONS?

It is good to be skeptical as a scientist. We are trained to be this way. The idea of new neurons in the adult brain was at first met with fierce skepticism, and some of that skepticism lingers today. But as more and more studies are published, more and more scientists have accepted the premise that new neurons are generated in humans throughout life.[14] The first human study was conducted in people who had throat cancer and, for diagnostic reasons, had been injected with a dye that labels new cells. After the cancer patients had died, the scientists were able to detect new neurons in their hippocampus.[15] More recently, a number of humans who died with dementia donated their brains to science. In this study, the number of new cells correlated with memory loss. In fact, the people who died with the severest dementia had fewer new neurons, while those with less of a deficit retained more of them.[16] Again, at this point in time, neuroscientists cannot "count" new neurons in people while they are alive, and until such a method exists, there will be some uncertainty regarding their presence and prevalence in humans. But to be clear, new neurons are not everything. They are just cells. To make them meaningful, we have to learn how to make the most of them.

PUTTING IT ALL TOGETHER

It has been well over twenty years since new neurons were rediscovered in the adult brain. What have we learned about them? Well, we know that they seem to arise in only a few brain regions, such as the hippocampus. We know that compared to older neurons, they are not great in number, but they do have one very special characteristic—they are new! As a result, they are especially responsive to what is happening both mentally and physically in this moment. As discussed, physical activities like aerobic exercise tend to increase the number of cells that are made, whereas mental activities that involve learning tend to increase the number of cells that survive. So, the question arises: What happens if you combine these two types of activities? Would it be better than doing just one of them alone? Germany's Gerd Kempermann and his research team attempted to answer this question with an experiment in mice.[17] For about a month, one group of mice lived in an "enriched environment," with new toys, other animals, interesting food, and so on. Another group of mice lived in a more secluded environment—but they could run on a running wheel. A third group had both the running wheel and the interesting environment. The mice in the latter group ran for ten days and then lived in the enriched environment for about a month. At the end of the experiment, animals that engaged in both activities retained more new neurons in their hippocampus than those that only ran or those that only lived in the enriched environment. So yes, for making and keeping new neurons, in mice at least, it is better to train the brain and the body together over time.

What about training the brain with mental and physical exercise at the same time? If you have ever watched the popular TV show *American Ninja Warrior,* you might recall the huge wooden wheels that the competitors try to stand on. As the wheel turns faster, it gets harder and harder for the warrior to remain on the wheel, and eventually most fall off into some water. But there are usually one or two who learn how to stay on long enough to get to the next challenge. This activity engages both the body and the brain in some serious skill learning. Neuroscientist Dani Curlik and I took this idea and built an apparatus for rats, during which they would learn to balance on a large rod that turned slowly at first and then faster and faster over trials.[18] As a result of learning this new skill, rats retained more new cells in their hippocampus. However, if the wheel turned very slowly, there was little to learn, and many of the new cells did not survive. So once again, new neurons in the hippocampus are more likely to survive if they are exposed to learning, as long as it is effortful. Now I know most people are not ninja warriors. But we can find ways to engage both our brains and our bodies in exercise.

"DESIRABLE DIFFICULTIES"

There is ample evidence that sustained training with mental and physical exercise changes the structure of the brain and may do so, at least in part, by increasing the numbers of new neurons in the hippocampus. But new neurons represent a small percentage of granule neurons in the hippocampus and even a much smaller percentage of neurons in the entire brain, most of which do not regenerate. All that said, new neurons make connections to older neurons, which make connections to still more neurons, and on and on. As I discussed, learning can enhance the anatomical connections between neurons, irrespective of whether they are new or not. So we may not know exactly the impact that neurogenesis in response to mental and physical training has on the brain—the human brain. But assuming it does matter, it would still take time and effort. Learning a new crossword puzzle once a month or running on a treadmill at the gym every few weeks is not likely to change the structure of the hippocampus by much, much less the size of the entire brain. To change our brain in a meaningful way takes concerted effort and sustained attention over time.

Let us now consider how we, as individuals, might actually increase the structure of our brains. In particular, we might consider engaging in *desirable difficulties*—a term used to describe mental activities that make learning more effortful.[19] One of these practices is spaced training, which simply means spreading the same information out over a greater length of time. This kind of training gives the brain more time to actually incorporate and then remember the material. Another desirable difficulty is self-testing, which involves quizzing yourself on the material without any clues. In school, we would associate this activity with flash cards. This practice works because it makes the brain work harder to retrieve the information. A third way to make the brain work harder is to interleave differing types of training at the same time. For example, instead of studying one topic at a time, we would study several topics simultaneously.

Some of these so-called desirable difficulties seem to enhance neurogenesis as well as learning. When laboratory rats are trained to learn a new task with trials spaced out over time, they tend to learn better and, as a result, retain more new neurons in their hippocampus.

When they are trained with different tasks one after another, this too seems to help keep the new cells alive, again as long as learning occurs.[20] But these are laboratory studies. In our real lives, it is often less than desirable to engage in effortful learning practices. Indeed, humans prefer massed training over spaced training because it is easier, even though it results in weaker memories. And few of us would attempt to acquire a new language while simultaneously learning how to play a new instrument and a new sport. Even if we were to make the effort to learn all three skills at one time, most of us would reach a plateau, after which any more training would feel aversive.[21] Nonetheless, training experiences that challenge our capacity for learning produce superior levels of performance. And when sustained over time, they likely enhance the structure of our brain along the way.

LEARNING TO SURVIVE

I want to circle back to the study about human depression and the hippocampus.[22] Recall that people who were depressed for long periods of time died with a smaller dentate gyrus and fewer neurons within it. But there is more to this story. Before the people had died, the researchers asked them to reflect back on traumas during their early life. In particular, they were asked to report whether they were ever separated from their parents by death or divorce or whether they had experienced sexual or physical abuse. In the people who had experienced these traumas as young children, the dentate gyrus was actually a bit larger and contained more neurons. Now it is important to realize that these cells were not necessarily new. They had been accumulating over many years, perhaps decades, and many of them were probably there since birth. And importantly, these particular participants were not necessarily depressed. Thus, their brains seemed to somehow compensate for the stress and trauma experienced while they were young. In general, these findings suggest that living with trauma memories does not necessarily lead to depression. Nor does it necessarily produce fewer neurons. Perhaps having extra neurons around might even make us more resilient. And vice versa.

There is a lot of talk these days about resilience. Resilience, resilience, resilience. What exactly is resilience? Generally, it is defined as the capacity to recover quickly from life's difficulties. Maybe this is what helped my dad. I told you about my dad and how he became very depressed when he first retired. Life had lost meaning for him. My mother was worried, and when she worried, she got busy. She made plans to get him out of his chair and out of the house. She eventually talked him into moving to California for the winter. There, they rented a condo along the beach, where he would walk down to the pier each morning to see what the fishermen were catching, often throwing in a line himself. She invited their friends to go out there, too. My brother and I lived close by and went to see them nearly every weekend. Looking back on it, I realize now that my mother was making a new and enriched environment for my dad, and luckily he knew he needed it. Were they resilient? Yes, I would say so. Both of my parents grew up poor and lost their parents early in life and in tragic ways. But somehow, they found each other, got married, had children, and traveled the world. And they were always learning. That was their thing. Did my father recover from his difficulties? In general, yes. He was never going to be seen skipping down the street. But he was content. He was resilient. And he lived to a very old age.

PART III

Preparing Our Brains for Everyday Trauma

8

Therapies for Stress and Trauma

I'm not afraid of storms, for I'm learning how to sail my ship.

—Louisa May Alcott, from *Little Women*

If you have had a serious trauma in your life, you might have sought help and know what kinds of therapies are available. But many do not seek help: some don't realize they need it, others don't know how to go about finding it, and many just flat out don't have the funds to pay for it. In this section, I am going to discuss some of the most accepted therapies for stress and trauma. Sadly, only a few are supported by scientific evidence. These therapies are referred to as *evidence-based* because there is evidence to support their efficacy.[1] They are verified through clinical trials, which most often compare a particular therapy to another therapy or to outcomes from people who have not received treatment. When practiced over time, they should produce both short- and long-term recovery effects across a range of symptoms. And importantly, these practices are conducted in a controlled environment with trained professionals; they must be safe. They also should be relatively easy to implement and scale up. Let me begin by describing the most accepted evidence-based therapy for trauma.

EXPOSING THE MEMORIES

Therapies for trauma often involve, in some way, exposure to aspects of an aversive memory, and as such they are generally referred to as *exposure therapies.* They emerged from the study of animal learning, which is simply the study of how animals (including humans) learn. Believe it or not, these studies go all the way back to the time of the great Ivan Pavlov (1849–1936). I am sure you have heard of Pavlov's dogs, and maybe you have heard of Pavlovian conditioning. When most people hear these terms, they think of reflexes—basic responses that are not very interesting. Or they think of dogs. I was once president of the Pavlovian Society, a society devoted to the study of learning. When I told my mom about my new position, she asked me if it had something to do with dogs! But seriously, the story of Pavlov and his dogs is quite interesting, and his observations made a huge impact in psychology and ultimately on the lives of people living with trauma.

Pavlov was born and raised in Russia, where he trained as a physiologist. Initially, he was interested in salivation and simply wanted to know how saliva gets released into the mouth in response to food. As the story goes, one of his technicians would walk down the hall each day to bring food to the dog. One day, Pavlov noticed that saliva was being released in the mouth of the dog even before the food arrived. How could that be? How could the food—which was still being carried down the hall—cause saliva to be released in the mouth of the dog? Pavlov reasoned that the dog must have anticipated the food. The dog had "learned" that the man in the white coat meant that food was coming. This seminal observation has been followed up by thousands upon thousands of studies about how we learn to associate events across time and anticipate the future.

This type of learning is called *Pavlovian conditioning,* and it is through these processes that we learn to become afraid. Take, for example, a woman who has been in a car accident. She is driving down the street, perhaps listening to the news or music, thinking about what she is going to do that day. All of a sudden, someone runs a stop sign, hits her broadside, and her car is totaled. Although she survives, she is forever changed. Now she is afraid to drive, maybe even afraid to be in a car, and if she has an extreme case, is afraid to go outside the house. In short, the woman has learned that driving the car is associated with the fear she felt when her car was hit by the other car. The next time she gets in the car to drive, it reminds her of what happened, and she feels afraid. The more she thinks about it, the more afraid she becomes. She starts to avoid driving. The fear may be irrational, but it still feels like fear. A memory connecting her car with a feeling of fear is now present in her brain.

How can we deal with this kind of memory in the brain? Can we get rid of it or at least get rid of the fear? Pavlov helped us out. He uncovered a process called *extinction*. To illustrate, let's go back to his dogs. The dog is now absolutely sure that when he sees the man in the white coat walking down the hall with the bowl, he is going to get some food, and he starts to salivate. But what happens if the dog sees the man, but the food never comes? Initially, the dog gets upset because he is expecting food. Then, once again, he sees the man, and once again, the food doesn't come. And again. And again. Eventually, the dog realizes the food is not coming. His brain has learned something new: the man in the white coat no longer predicts the arrival of food—and the dog stops salivating. This learning process is known as *extinction learning*. Importantly, the dog did *not* forget about the food or its connection to the man in the white coat. He has simply learned to stop responding. The memory connecting the man with the food is still in his brain. But he has a new memory in which the man is not connected to the food.

Now back to the woman who is traumatized after her car accident. How in the world is she going to get out of her house and back into her car? One way is to start training her brain to learn that all cars do not cause accidents. Perhaps she could start by watching videos of happy people driving cars with the top down on sunny days—or even photographs of herself in better days in her own car. She might go with her friend on a car ride around the neighborhood or just go sit in the car in the driveway. Then she might drive herself around the neighborhood. Eventually, she might even drive down the exact street where she had the accident. All these experiences are called *exposures,* and just like Pavlov's dog, she is creating new memories, memories where cars are not connected to accidents. Her fear of driving, even of driving near the accident scene, should start to dissipate. She is exposing herself to aspects of the old memory, while creating new, accident-free memories, so she becomes less fearful through repeated exposure. It is important to note that people should not be doing exposure therapy on their own. This type of therapy, known as *prolonged exposure,* or PE, is provided by a trained professional, usually a clinical psychologist or psychiatrist.

LINGERING WITH THE MEMORY

Prolonged exposure therapy was developed by Edna Foa, renowned psychologist and professor at the University of Pennsylvania.[2] I met her once at a conference at Yale. The speakers were all staying together in a quaint bed-and-breakfast. As we were getting settled, Dr. Foa came flying into the room with a big scarf and flowing dress, filling up the room with her personality. In the 1980s, Dr. Foa theorized that PTSD was caused by trauma memories in the brain and wondered whether people, like Pavlov's dogs, could learn to extinguish the memories. As the person with trauma memories is exposed over and over again to reminders of the event, they start to replace the feeling of fear with a less fearful response. In the case of the woman in the car accident, she is replacing her memory of the car and how she felt at the time with a new memory in which she is not afraid of the car. She has learned something new.

When someone signs up for PE, they will usually attend about ten or twelve sessions, each lasting about an hour and a half. In the first few sessions, the person is given information. The therapist might discuss common responses to trauma and suggest simple tools to reduce stress such as breathing exercises (although not to be used during the actual exposures). In the next few sessions, the therapist will purposely begin to expose the client to the fear memory: the client is asked to think about what happened, say what happened out loud, sometimes write down what happened, and sometimes, when possible and not considered too traumatic, visit the place where it happened. The last few sessions of therapy are more geared toward the future. The client will review her progress and figure out strategies to prevent relapse. In particular, she will consider what she will do when confronted with triggers—reminders of the fear. Now to be clear, this is not a straightforward process. Exposing oneself to the memory requires that someone is open to telling the details and then recounting and recording them in some way—either verbally or in written form. But because the event itself has passed, the recovery of those memories is not easy, nor is it obvious how to do it in real time. The process must be methodical; this is why trained professionals are necessary.

How does exposure therapy actually work in the real world? It is not easy, and you can see why. In one study, women who had been raped and were experiencing many symptoms of PTSD attended therapy twice a week for about an hour each session. During the initial sessions, the client lists situations or reminders of the trauma. Then in each subsequent session, she is asked to get closer and closer to the actual memory by imagining the traumatic event. She is then asked to describe it out loud in detail as it happened. She is asked to do this in the present tense—to make it seem like it is happening again in the present moment. As you can imagine, this process can be painful and disturbing because most people do not want to revisit, much less relive, the memory. They do not want to reexperience the fear. But fear is the "active" part of the therapy. Without reliving the fear, it will not extinguish, at least according to the theory. After this, then what? Well, the process becomes even more difficult. Now the therapy session is taped, and the person is asked to go home and listen to the imaginal exposures captured on the tape—at least once a day. They are also asked to engage in about forty-five minutes of behavioral exposures on their own each day. In this case, the therapist does not typically recommend that a client actually visit the scene or look at pictures of the perpetrator. But even so, it is a difficult process. Difficult yet effective.

There are variations to this type of therapy—but all of them depend on reactivating aspects of the fear memory—over and over and over again and always in a safe context. Indeed, the therapy is called *prolonged exposure therapy* because the fear must be reactivated again and again in order to extinguish the memory. Memories are stubborn things, and they do not go away easily. Remember Pavlov's dogs? You would think that after a few times of not getting food, the dog would forget to salivate when they saw the man with the bowl. No way. It took trials upon trials of training to stop the salivation. And you know what? Even after the dog quit salivating in response to the man, the dog needed to experience the man with the food only one more time, and salivation started up again—often more than ever. Just like the dog, people learn quickly to go back to the most salient memory. Consider the woman in the car accident who has now recovered and is calmly driving all around town. Suddenly, she sees someone run a red light. The fear memory comes rushing back, like it was sitting there just under her skin waiting to burst out.

Time does not always help. I had a young woman in one of my classes many years ago. She came up to me after class, clearly very upset and wanting to talk. Back in high school, she had gone to a house party with her sister. A young man she did not know took her into a bedroom and raped her. She never told a soul—not even her sister. She went through the rest of high school avoiding her rapist and even participated in about ten sessions of exposure therapy at her college health center. Then one day, she was home for the holidays, shopping and walking through the mall. She turned a corner, and there he was, standing there looking in a shop window. The second she saw him, she froze. She couldn't move. Her friends did not know what was wrong with her. She went home immediately and got physically sick with fear. Years after the trauma, seeing his face brought back the memory, along with the fear she felt in her body. Fear memories are especially tough. They cannot be erased. It is difficult enough to get them to fade into the background. Many people drop out of exposure therapy—as many as one in three don't come back. Exposure does help, but it also takes work, bravery, and determination, and sometimes the pain is, completely understandably, too much to bear.

PROCESSING THE MEMORY

Somewhat related to PE is a newer therapy known as *cognitive processing therapy*, or CPT. It was initially developed by Duke psychologist Patricia Resick for survivors of sexual assault but is now recommended for people with many types of trauma histories.[3] Like PE, CPT is evidence-based. And like PE, the client focuses on a traumatic event in their life, usually by recalling it in detail, verbally but sometimes in writing. But in addition, this program focuses on *beliefs*. As discussed, many people blame themselves for what happened in the past, and these beliefs can perpetuate the symptoms of trauma, such as anxiety and depression. As Tara, a young woman from Washington, D.C., relayed after coming home from work to find her roommate in excruciating pain. At first, she did not take it seriously and sat down to watch some television. But then she heard her roommate crying hysterically in the bathroom and called for help.

> I felt guilty for not calling the ambulance sooner, and am always wondering whether I acted fast enough to get her help to minimize the damage. I keep finding myself thinking about the quick turn of events and wondering what would have happened if I had not come home when I did.

After a few information sessions about PTSD and trauma-related symptoms, the client is asked to provide a detailed account of the events surrounding the trauma and then to consider why they think they occurred. Thus, during CPT the client revisits the trauma memory, and the therapist guides the client in exploring what they believed before the trauma and what they believe now. The therapist encourages the client to acknowledge this transition in beliefs, posing questions in a somewhat Socratic method. For example, the client may have felt safe before the trauma, but now they do not believe they are safe. Or they might think it was their fault somehow. The therapist would raise questions about these beliefs and, as a follow-up, would have the client complete daily worksheets at home.

My friend Melanie, whom I discussed earlier in this book, may believe she could have prevented her father's friend from abusing her, even though she was a mere child at the time. During CPT, she would hopefully come to realize that what happened was in no way her fault and there was nothing she could have done to stop him. This process leads to a decrease in guilt and blame, as well as developing more trust in others going forward. In essence, the client is learning new mental skills that help them recognize and reevaluate thinking patterns—including negative thoughts and illogical beliefs associated with the event. The client learns to accept the fear, and as a result, it becomes less powerful. The goal is not necessarily to change the memories of the past but to move more fully into the present with a greater sense of control about the future. In practice, this program is similar to PE, with about twelve weekly or semiweekly sessions, beginning with information and then transitioning into the actual memory work. Similar to PE, there is "homework," such as cognitive worksheets and narratives to write down, with the goal of learning new mental skills outside the sessions. The homework is not as intensive as during PE, but it does require considerable time and effort. As with most things we do in life, the more effort put into them, the better the outcomes.

WHICH TECHNIQUES ARE BEST?

Of course, the big question is: Do trauma-focused therapies work, and is one better than the other? One study compared and contrasted the responses to two types of therapy—PE and CPT—in women who had been raped, anywhere between three months and thirty years earlier.[4] Approximately 170 women were admitted into the study. Nearly half had been raped twice in their lives and many were sexually abused as children, and all were diagnosed with PTSD. Thus, they were experiencing intrusive thoughts, anxiety, and arousal, and were often reliving the trauma. About one in four women did not show up for the first session or dropped out of the study, regardless of the therapy type, with about forty women completing each program. The program included two sessions per week for about six weeks, and all the therapists were trained to follow the format described by Drs. Foa and Resick, the two creators of PE and CPT, respectively. After treatment, most of the participants no longer had PTSD and maintained their progress even nine months later. Participation in the program even exceeded the clients' expectations. Also, they reported many fewer symptoms of depression. The amount of guilt was reduced in both treatment groups, with CPT being a bit better for some of these particular thought processes. And both therapies helped women regardless of when their most recent trauma occurred, as far back as thirty years! Given these results, anyone who thinks it is "too late" to get help, because their trauma happened so long ago, should reconsider.

How does imaginal exposure to trauma memories help people thrive in the present? Trauma memories are intense and difficult to forget, but they are often disorganized. People either don't remember parts of what happened because they don't want to or because the memories were not encoded well during the actual event. Remember that the brain needs time to respond to a trauma and often doesn't become completely activated until the event itself is nearly over. And then later in life, the person often rehearses the memories over and over again—even though the event itself may not be well remembered. Some of these thoughts may become more and more repetitive, as they expand out from the event itself into related thoughts that focus more on self-blame and guilt for what happened. It is as if the person is trying to understand what happened but can't. According to Dr. Lily Brown, clinical scientist at the University of Pennsylvania and an expert on therapies for PTSD, "exposure therapy helps the memories become more organized." I like this way of thinking about therapies for trauma. Most of us don't like revisiting traumatic memories, but in a safe context doing so repeatedly can help organize the memories and promote healing.

SEEING ANOTHER PERSPECTIVE

Since I mentioned Pavlov, I might as well mention Freud, the founder of psychoanalysis and the so-called talking cure. Sigmund Freud was born less than ten years after Pavlov and, like Pavlov, was not a trained psychologist. Rather, his training was in medicine, during which he began to focus his considerable energy on trauma and its memories. In particular, Freud considered the possibility that many of our most traumatic memories are not readily available for us to think about—because they are stored in our unconscious. He practiced talk therapy, along with dream analysis, as a means for resurrecting those memories and bringing them into conscious awareness. Most modern-day psychologists have moved away from psychoanalysis, but many of Freud's ideas remain, as well as various renditions of his practices. When I first started teaching psychology back in the 1990s, I rarely lectured about Freud and certainly did not give much credence to psychoanalysis or to psychodynamic approaches in general. But through the years, I have grown to appreciate his teachings and some of the alternative therapies that have emerged from them.

Take, for example, the empty chair technique. To begin, the therapist brings out two chairs. Each chair is given an identity—either a person or a trait or a viewpoint (say, two people in a disagreement). Then the client is asked to speak from one chair about their trauma or concerns. When the dialogue reveals the other person or viewpoint, the client is asked to move to the other chair and speak from that viewpoint. And back and forth. This process, of going back and forth, helps the client integrate the differing views and opinions. This technique is complex and depends on a healthy relationship with a trained therapist because strong emotions are likely to appear as the client progressively feels freer to reenact the past. But with time, the client begins to view the trauma in a larger, more coherent context. It helps organize the memories.

Let me tell you a story I heard once about a man named Mark. He had a difficult childhood, living with few of the resources many of us take for granted, such as food and shelter. He also had a tumultuous relationship with his older brother that bordered on abusive. As an adult, Mark still harbored hard feelings toward his brother and could not find a way to let go of the past. He did, however, find his way into therapy and into the empty chair. As Mark learned to express his feelings to his brother, who was imagined to be sitting in the empty chair, he began to realize that he could stand up to the bullying and to his brother. But as Mark moved into the other chair, he was also able to see things from his brother's point of view. He was able to put the memories of his childhood not only in order but in context as well. His brother, after all, had grown up under similar impoverished conditions and was often put in charge of the family while their mother was working. After going back and forth between chairs within and across sessions, Mark learned to view his past within a more coherent and organized narrative. Along the way, he learned to feel empathy for both himself and his brother.

HEALING THROUGH THE BODY

Around the turn of the century, I was teaching a graduate seminar for students studying for their doctorates, most of them in clinical psychology. And most were fully committed to modern approaches, especially prolonged exposure therapy and cognitive behavioral therapy. But they also told me, in no uncertain terms, that they wanted to know more about a new wave of therapies that focus not only on thoughts and behaviors but on feelings in the body as well. The most well-known practice at that time, and still today, is *mindfulness-based stress reduction*—known as MBSR. This program was developed by Jon Kabat-Zinn, a prolific writer and clinical researcher at the University of Massachusetts.[5] As a young man, Kabat-Zinn had studied meditation techniques and then adapted the practices for medical settings, stripping them of most religious or cultural appendages. The training program relies heavily on both sitting and walking meditation, as well as yoga and other activities that enhance body awareness. MBSR was not developed for trauma, although it can lessen some symptoms, such as anxiety. Also, MBSR is not a therapy, per se, and is therefore not done in the presence of a therapist but rather with a facilitator, who participates in and leads the practices. And the program was not designed to address any particular disorder or symptom but rather to reduce stress, which everyone feels and can use less of. This program does take time—it is usually practiced nearly every day over about eight weeks. It also takes dedication because meditation is difficult to practice consistently. But anyone can do it, and people like doing it. And if people like doing something, they are more likely to actually do it.

Since then, therapies that focus on meditation and the body have proliferated.[6] Especially popular are those that combine traditional cognitive approaches aimed at recognizing and reevaluating maladaptive thoughts and beliefs with those that rely on meditation, yoga, or body-centered techniques. Two examples are *mindfulness-based cognitive therapy* (MBCT) and *dialectical behavior therapy* (DBT). I encourage people to shop around and explore the many therapies that exist. It is natural to gravitate toward those that are most appealing or seem easiest. But it is also important to consider what actually works for your individual situation.

LESSENING THE STRENGTH

It would be nice if there were a pill that we could take to get rid of unwanted thoughts and memories, but so far there is none. However, there are medications that help reduce the strength of symptoms associated with trauma memories and, in particular, PTSD. For example, many people with PTSD experience symptoms of anxiety, and as such, they are often prescribed "antianxiety" drugs. These medications are more accurately referred to as *benzodiazepines,* and they act to increase inhibition by increasing GABA, which is the major inhibitory neurotransmitter in the brain. The transmitter reduces neuronal excitation. You can imagine why these medications might reduce anxiety—but also why they influence other thought processes and behaviors as well. For example, some benzodiazepines relax the muscles, which is why they are often prescribed for muscle injuries. Also, people can build up a tolerance over time, meaning that more of the drug is necessary to get the same effect. They are especially dangerous if paired with other drugs that inhibit brain activity, such as alcohol or opiates. However, when taken under the supervision of a physician, they can be valuable.

In particular, benzodiazepines are effective for relieving panic during a panic attack, a debilitating symptom in some people with PTSD. In this case a medical doctor might instruct the person to take a pill when they start to feel the attack coming on. This makes sense, and in some cases they are prescribed more often. Johnathan, a young man from Canada, was diagnosed with generalized anxiety disorder, including panic attacks, and was initially hesitant but changed his mind:

> I was not convinced that the medications would do anything for me, but I knew my anxiety was real. Most people don't have panic attacks every other day and rip up their own skin. Once I started on the meds, I noticed a change. At first, I brushed it off because I am very anti-drugs and skeptical of anything that claims to "fix" you. But as the days wore on, I realized that I was no longer anxious about everything in the world.

People with trauma symptoms are sometimes prescribed "antidepressants," which are more accurately referred to as *selective serotonin reuptake inhibitors*—or SSRIs. Serotonin is a neurotransmitter that passes from one neuron to the next by traveling across a synapse and binding to a receptor. Under normal circumstances, extra serotonin is taken up and stored, but these medications release it back into the synapse. They can also change the anatomy of the brain, even making new neurons in the hippocampus.[7] And not all "antidepressant" drugs work on the same neurotransmitter system. For example, some target norepinephrine, which is released along with epinephrine from the adrenal glands as well as from the brain. These medications can give people a bit more energy, which is often desirable for those suffering with feelings of depression.

Whether psychotropic medications help with trauma symptoms, per se, is hotly debated in the field of psychology, albeit less so in psychiatry. Some people respond very well and others not as much. And some people respond positively to taking medications regardless of what the medication is doing to the brain. This effect, known as the *placebo effect*, is still mediated by the brain, just not as a result of the pharmaceutical itself. Regardless, these options must be considered with the supervision and advice of a physician and their medical staff.

NOT ALWAYS READY TO REMEMBER

There is no magic cure for stress and trauma. Some approaches work better than others, and some work for some people and not others. Some approaches focus on thoughts, beliefs, and memories, such as prolonged exposure and cognitive processing therapies, whereas others focus more on the body, such as mindfulness-based stress reduction. Still others work by changing brain chemistry, such as antidepressants and antianxiety medications. I am not a psychiatrist who prescribes medications, and I am therefore admittedly biased toward therapies that focus on learning new skills that can help someone recognize thought patterns while learning new ways of interacting with memories, as well as the feelings those memories elicit. But when people need help, we must consider all approaches.[8] And there are always new interventions to be discovered. For example, digital applications and internet therapies are now commonplace as the need for them expanded during the pandemic. Indeed, there are too many options to discuss them all here, and my goal is not to do so. Each person must find the path that works for them without becoming overwhelmed by all the choices. As psychiatrist Dr. Christopher Fairburn often says, "Simpler procedures are preferred over more complex ones; it is better to do a few things well rather than many things badly."[9]

Psychological practices for trauma, in particular, must be recommended with caution and compassion. At one point while writing this book, I considered titling it *Ready to Remember* because so much of our experience with trauma relates back to memories, but then I ran it by one of my best friends, who has had a lot of trauma in her life. She recoiled at my suggestion. She said there are some things she is *not* ready to remember. It might be easy for some of us to tell people to forget what happened to them in the past. Or worse, we might tell them that it is good to think about it even more than they already do, without fully realizing how painful that may be. Clearly, I did not end up using that title, but I use the story behind it to remind myself how others feel about revisiting their own memories.

9

MAP *Train My Brain:*

A "Mental and Physical" Training Program

Learning about neurogenesis has profoundly affected the way I live my life. Now I realize that the health of my brain isn't just the product of my past; it reflects the choices I make and the experiences I have today. This is extremely good news.

—Email from Francesco

The other day, I was scrolling on Twitter when I saw a picture of an old antebellum-style building that felt very familiar. I immediately stopped to see what it was. Sure enough, this building was once a mental institution where I worked during college. Every week, I would drive there to lead an art class followed immediately by an exercise class set to music. I don't remember much from those days, but I do remember that some days the patients wanted to participate and, other days, not so much. Sometimes they were happy to see me and, other times, not so much. And some days I would leave feeling like I had helped them somehow and they would get better, while other days, I left feeling totally incompetent and helpless. As an undergraduate, I knew next to nothing about mental health and even less about the brain. I thought I could heal people with the sheer force of my personality!

Since then, I have learned much about the brain and how clinicians go about trying to help treat people who are suffering with the everyday traumas of life. But sometimes, I wonder how far we have really come. It is still very difficult for people to find help or even know how to go about asking for it. And even when they know how, it is often out of reach—either because it is too expensive or too time-consuming. And I have been surprised to learn how many people are actually reluctant to try therapy or continue once they try. As one person described her experience in therapy:

> I remember sitting in the waiting room being hyperaware of everything. There was a piano playing softly from a speaker; however, rather than being calming, it made me feel more uncomfortable and frightened. When I was finally called in to sit with the therapist, my heart began racing and my arms felt sweaty, my legs were shaking uncontrollably, and I couldn't stop fidgeting. I didn't want to talk about what I was going through, especially not to a stranger. I kept thinking about just getting up and leaving without even saying goodbye. After I left the office, I kept picturing myself sitting in her office. I felt trapped in the moment and could not imagine myself having to go back again.

Please do not get me wrong. I am a therapy advocate, especially forms that are evidence-based and cognitive in nature. However, for some people, access can be difficult, and then staying with it has its own set of problems, as described for exposure therapy. But above all else, most people do not seek out clinical help until they feel like they need it or something traumatic has happened. Perhaps we could help ourselves beforehand by engaging in activities that make us more resilient, helping us get through the bad times with less stress and feelings of despair. And perhaps we could even find a way to ruminate less in general—so that when the traumas do arise, we will be less inclined to make yet more memories of them in our brains. We need a program that can help us deal with problems that are already present and those that are yet to come. We need a program that anyone can do, regardless of income or gender or race or age or even need. With these goals in mind, I have developed such a program. It is called *MAP Train My Brain*—or MAP Training for short.

The elements of the program are actually not that different from the classes I used to lead as an undergraduate working in psychology, at least with respect to combining "mental and physical" or "MAP" activities. But this program capitalizes on what we know now about the brain—its capacity for change and its capacity to learn. Importantly, MAP Training is not a therapy and is not meant to replace traditional therapies or medications. Rather, it is a brain fitness program that we can all use to help us live with the everyday stress and trauma of life.

THE MENTAL PART OF MAP TRAINING

As you may recall, my lab discovered that new neurons in the adult brain could be rescued from death by learning. But not all forms of learning were equally effective. To actually increase cell survival, the learning process had to be effortful—it had to be difficult to learn.[1] When I first started thinking about how to translate this information into a fitness program for humans, I considered the most difficult of all brain training games, known as the *N-back*. During this task, you are asked to remember a series of numbers or letters or pictures presented in rapid order. Then starting with the most recent item, you go back in order and try to recall items further and further back. If N=2, then you must try to recall items going back two at a time, and then with N=3, you must try to recall items that are three back, and then four, and then it becomes very difficult. Obviously, this brain training task requires a great deal of concentration, and with training, you will get better. But sadly, this "game" is not very appealing. I asked my students to do it. They tried it for a few days and quit. I tried it for a few days and quit. It was too hard and not interesting. Unfortunately, many brain games are like this.

Then a friend suggested meditation. At first, I rolled my eyes. Looking back, I realize I was confused. I thought of meditation as too "soft" and unscientific for my brain training program, and at the same time, it seemed too difficult for people to actually do—myself included. Mostly, I had no idea what it really was. So, I went exploring and discovered for myself that, yes, meditation is difficult, and it does require effort. But it is also extremely interesting and a true learning experience. Initially, I was trained in a form of meditation referred to as *zazen*. In clinical circles, it is called *focused-attention (FA) meditation* because you focus your attention on one thing—usually the breath. During this practice, you are instructed to focus your attention on your breath and then start counting each breath. When the mind starts to wander off, you are instructed to recognize that you have lost count and return your attention to the breath, beginning your counting again at one. The breath is useful because it is always with us and it is always changing. It is also useful because it is not inherently that interesting, and therefore you really do need to focus your attention. It would be easier to focus your attention on watching a good movie or riveting TV show or what might be trending on Twitter. But this would not be meditation. And it would certainly not be brain training.

SITTING IN SILENCE

Sometimes people think meditation is about learning *not* to think. But this is not really the point or even possible. Rather, while sitting in silence, our brains are learning a new mental skill. We are learning to see our thoughts coming and going, without necessarily engaging with their content. If we do engage, we are learning to understand that we have chosen to do so. It is also an effective way to learn what we "normally" spend our time thinking about and how often. Most of us spend most of our time thinking about ourselves. We think about how we can survive—*How can we find food and shelter and love? How can we stay safe?*—but we also engage in loads of everyday thoughts about what we are going to do later in the day and who we might see or want to see and miss or where we might go or wish we could go. Many of these thoughts are repetitive—we repeat them over and over. Some are inspiring and informative, but many are not. One woman told me that she was going on a trip to South Africa and could not stop packing her suitcase in her mind, even though she had already decided long ago what to take on her trip. Many of our day-to-day thoughts are depressing or full of worry, while others are brooding and full of blame—those ruminative thoughts that I have discussed at length. Many are connected up with memories from the past that we would just as soon not go over again. During meditation, we learn that we do not need to follow all our thoughts, or even most of them. We learn to let some of them come and go and then come again and go. As Zen teacher Shunryu Suzuki once said, "Leave your front door and back door open. Allow your thoughts to come and go. Just don't serve them tea."[2]

Think of your mind as a muddy lake that, as you meditate on your breath, starts to become clearer. Indeed, with practice, you might "see" the beginning of a thought as it is forming in your brain, as it becomes real—maybe turning into a word or a sentence or a story. You might see it trying to bring up an old memory and integrating that memory into what you were already thinking about. You might start to feel the memory bring up an old feeling in the rest of the body. You might even begin to see a thought as it ends. Apparently, some expert meditators can discern the space between each thought. I'm still working on that! Importantly, there is no real goal during meditation, and it is certainly not a competition. It is simply an opportunity to learn more about your own brain and the thoughts and memories that it is generating, millisecond by millisecond. As we learn more about these ongoing mental processes, we learn how to live more peacefully alongside them in our everyday lives, even in the face of stress and trauma.

WALKING SLOWLY

When I first learned to meditate, I learned two forms of practice. The first is the sitting meditation practice that I briefly described, during which you sit upright in silence and count your breaths. The second practice I learned is referred to as *walking meditation,* during which you walk very, very slowly. While walking, you direct your attention into your feet, much like we did with the breath while sitting. But this time, you try to feel your feet and how they are experiencing the act of walking, as you step from one foot onto the other, feeling your weight transfer from one side to the other. During this process, we are becoming more aware of how our bodies are moving through space and how each moment is different from the next, even doing something as simple and automatic as walking. But more important, we are training our brains to pay attention. As we walk slowly going nowhere, we try to stay focused on our feet, but like our breaths, it is not that interesting, and before we know it, our minds are off again, wandering around and wondering what we would rather be doing or thinking about what happened earlier in the day or ten years ago. And as we notice that we are no longer focused on the feet, we simply bring our feet back into focus. With practice, we might learn to walk like a camel, as the writer Henry David Thoreau claimed he could do.

One time I was meditating with a group of experienced meditators. We had been meditating for several days, hour after hour in a sitting position, facing a wall, no less. Finally, the guide told us that we would be walking outside. I was excited to hear this because it was a beautiful sunny day up in the Catskill Mountains, but then he told us not to get excited. What? Why not? After all, we were going out into nature, where we could explore all the wonders of life. *Why is the sky so blue? Where is that river flowing? What is ahead over the hill? I am so lucky to be here. To be alive.* He told us to just walk, to not let our minds run away from our feet. Gosh, I wanted to stay excited. But after a while, I was relieved, relieved to just walk and focus my mind on that one task.

As with sitting meditation, the skills learned during walking meditation can go well beyond your living room or the Catskill Mountains. One of my close friends, Dr. Roberta Diaz Brinton, heads up a brain institute to help women with Alzheimer's disease live better lives while experiencing less memory loss. Nearly thirty years ago, she told me something that I still remember today. I was talking about all these great ideas I had and all the things I was going to do with my great ideas. She stopped me midsentence and said, "Tiny steps for tiny feet." We are always running. Running here, running there, trying to do this and that and get more and more ahead. And sometimes (most times) feeling like we are getting nowhere. But we can become conscious of where we are and the steps we are taking. We can become aware of our tiny little feet.

THE PHYSICAL PART OF MAP TRAINING

Now for the physical part of MAP Training. The choice here was easy—aerobic exercise. As discussed, aerobic exercise can increase neurogenesis in the hippocampus, but it has many other benefits for the brain. The brain uses up to 20 percent of the oxygen we breathe, even though it makes up only about 2 percent of our total body weight. Why does the brain need so much oxygen? Neurons. They need oxygen to fire, and they are firing all the time. Much of the energy is used to generate electrical current at the synapse—the connections between neurons.[3] The brain also uses oxygen to create connections and to create new cells, even the new neurons. In one study, mice were given the opportunity to run on a running wheel as much as they wanted, and they love to run. Within just a few days, the hippocampus contained more blood vessels, especially in the region where the new neurons are generated. Not only that, new blood vessels showed up before the new neurons did, suggesting that neurogenesis depends on their presence.[4] But again, these changes are not confined to the hippocampus. The entire brain depends on lots of oxygen to do what it does—think and learn, remember and forget. In fact, one study reported that the hippocampus uses more oxygen to forget than it does to remember, but only as long as the forgetting is intentional.[5]

The message is clear: aerobic exercise is good for the brain, and because it is good for the brain, it is good for mental health. One study was especially impressive, collecting data from more than a million people.[6] Those who engaged in exercise reported significantly fewer days of poor mental health than those who did not exercise. The most effective types of training were cycling, running, and team sports—all aerobic in nature. Aerobic exercise seems to even be good for the size of the brain. One study reported that people who exercised and had greater cardiovascular fitness had a larger hippocampus.[7] Whether or not a bigger hippocampus can prevent dementia or depression is debatable, but I would rather have a big one than not.

Scientific studies on aerobic exercise are many, and some of them indicate benefits for learning and memory as well. One study recruited middle-aged men and women with mild cognitive impairments to engage in high-intensity aerobic exercise four days a week for six months. Other groups of men and women stretched for the same amount of time. In general, both sexes benefited, but the women made especially impressive gains in cognitive flexibility and information processing speed. Moreover, levels of the stress hormone cortisol were reduced in women who exercised aerobically.[8] And it seems that combining mental with physical activity may be especially beneficial. A large group of middle-aged women were followed for more than forty years.[9] The participants who engaged in intellectual and artistic pursuits in addition to regular physical activity were less likely to suffer cognitive loss as older adults. Now to be clear, all people lose some memory function with age. This is just a fact. But these data suggest that engaging in both mental and physical training over significant periods of time may mitigate some of the loss.

To be aerobic, the heart must be beating fast. To find your zone, simply subtract your age from 220 and then multiply that number by 0.6. This number approximates the lower limit of your aerobic zone. Then multiply the same number by 0.8. This number reflects your upper limit—the highest you should let your heart rate reach during the training. For most adults, a rate greater than 100 beats per minute lies within the aerobic zone. Getting the heart rate up requires effort but can be accomplished in a variety of ways—running, jumping, spinning, high-intensity dancing, whatever it takes. The best way is the way that works for you. After years of doing this myself and trying to motivate others to do it, I have learned that people have very particular tastes in exercise. Some people love to run, whereas others despise it. I prefer aerobic dance exercise, but others despise it. Spinning is popular these days. For the *MAP Train My Brain* program, it does not matter *how* you exercise as long as it is aerobic. But there are tricks. For example, putting your hands over your head while doing any of these activities will elevate the heart rate. After all, you need to get oxygen up into the arms, which are now above the heart. Same with the feet. Kicking them up into the air requires much more oxygen than shuffling them along the street. No matter what type of exercise you decide to do, you should be sweating!

MENTAL AND PHYSICAL TRAINING TOGETHER

That's it! The *MAP Train My Brain* program combines mental and physical training—sitting and walking meditation, followed by aerobic exercise. My goal was to create a program that helps people learn more about their own brains while learning to think less often about the past. I wanted to create a program that would benefit anyone—no matter their age or race or gender or income. I wanted to offer a program that would not require a lot of time or money or fancy equipment. So, let's do it. Let's MAP Train our brains. But before you begin, please take a snapshot of how you feel in this moment. Get out a pen and a piece of paper and number it from one to twelve. Then answer the following questions, noting your answers with the letters on the paper. Wait until you are finished to tally your score.

MAP Health Survey

Think about your life in general, and rate each statement from hardly ever *to* all the time. *Ignore the numbers for now.*

1. I like hanging out with other people and can't wait to do it again.

   a. hardly ever (4)
   b. sometimes (3)
   c. often (2)
   d. all the time (1)

2. I think about the things that I have done wrong.

   a. hardly ever (1)
   b. sometimes (2)
   c. often (3)
   d. all the time (4)

3. I wonder why I can't remember what happened in the past.

   a. hardly ever (1)
   b. sometimes (2)
   c. often (3)
   d. all the time (4)

4. I sleep well and feel pretty rested for my age.

   a. hardly ever (4)
   b. sometimes (3)
   c. often (2)
   d. all the time (1)

5. I have enough energy for doing things in life that I like to do.

   a. hardly ever (4)
   b. sometimes (3)
   c. often (2)
   d. all the time (1)

6. I feel like I am to blame for the problems in my life.

    a. hardly ever (1)
    b. sometimes (2)
    c. often (3)
    d. all the time (4)

7. I think that my life would be better if I had more time alone to think.

    a. hardly ever (1)
    b. sometimes (2)
    c. often (3)
    d. all the time (4)

8. I can't seem to relax my body.

    a. hardly ever (1)
    b. sometimes (2)
    c. often (3)
    d. all the time (4)

9. When I think about going out in the world, it makes me want to stay home.

    a. hardly ever (1)
    b. sometimes (2)
    c. often (3)
    d. all the time (4)

10. I am excited to learn new things.

    a. hardly ever (4)
    b. sometimes (3)
    c. often (2)
    d. all the time (1)

11. I find myself thinking about myself in the past.

    a. hardly ever (1)
    b. sometimes (2)
    c. often (3)
    d. all the time (4)

12. I feel good about how things are going in my life.

    a. hardly ever (4)
    b. sometimes (3)
    c. often (2)
    d. all the time (1)

> **MAP Health Score Calculation**
>
> To find your MAP Health Survey score, simply add up the the numbers in parentheses next to your choices. The lowest possible score is 12, and the highest possible is 48. In general, a higher score suggests you are feeling more stress-related thoughts and feelings than a lower score. But please don't worry too much about the total score or what it means just yet. This is not a pass-fail test but rather a simple way to get an idea of what you are thinking about and how you are feeling in this moment. You can take the survey again at any time—but I recommend waiting until you have completed a MAP Training session at least once a week for six weeks.

HOW TO MAP TRAIN YOUR BRAIN

Before we begin, there are a few more things to consider. If you have health concerns, make sure to check in with your physician or medical professional. Also remember that this program does not replace other treatments and therapies. Once you are ready, change into some comfortable clothes—something you can move around in and not worry about getting sweaty! Have some water on hand and some running shoes, what we used to call *tennis shoes.* Find a quiet place without a lot of distractions, and most definitely silence the ringer on your phone. You will need a timer, however. The one on your phone or an old-fashioned egg timer is fine.

MENTAL TRAINING: THIRTY MINUTES OF SITTING AND WALKING MEDITATION

- First, we are going to sit for twenty minutes in meditation. If you can, find a hard pillow or cushion and sit down on the ground on top of it. In meditation circles, people use what is called a *zafu cushion,* which is like a pillow but hard and round. They are good to use, but if you don't have one, just use an old cushion from your couch. Sit down on the front edge of the cushion. You want to be sitting straight up, not leaning back against the wall or onto another cushion. Sit as erect as possible, with your legs crossed in front of you. If you have bad knees or this is impossible for you to do, then you can sit in a straight-back chair. Place your backside on the edge of the chair with your feet close together on the floor. Again, if possible, try not to lean back but rather sit upright and erect. We need to be alert and awake to meditate. This practice is not about relaxation. We are training our brains.

Set your timer for twenty minutes.

- Now for your arms. You want to have your arms loosely at your sides with your hands in your lap. Place the left hand on top of the right one with your thumbs touching together ever so lightly, as if you could place a piece of paper between them. This posture is meant to be comfortable but again not too comfortable. The posture also has symbolic meaning—as if your right hand were holding your heart.

- Now for your eyes. You can either close your eyes completely or close them halfway. If you are one of those people who falls asleep immediately when they close their eyes, then definitely keep them open a bit with the focus about three feet in front of your face looking toward the floor. Just don't be looking around the room or at yourself in the mirror!

- Now for your breath—and your brain. Take a deep breath. Feel the air go in and out. Do it again. Feel the air go in and out. Focus your attention on the actual air going in and out. Don't breathe too fast. Just breathe normally. After doing this a few times, start to notice the space between the out-breath and the in-breath. This little space is often called the *little death*, because you are neither breathing in nor breathing out. It is nonetheless a space in time in your body that you can always find. This space is going to be the focus of our attention. So again, notice the space. Now count it, starting at one. And count all the way to a thousand! Of course, you probably won't count all the way to a thousand. Rarely does anyone. The point is to just recognize when you lose count and when you do, to return your attention to the space between the out-breath and the in-breath starting again at one. Try not to worry about the counting. This is not a competition. Counting is just a good way to keep yourself (and your brain) focused on one thing at a time. After trying it for the first time, one participant told me, "I had never meditated before, so I was curious about that part. Although I tried to count numbers as instructed, I frequently lost count and needed to start over. Sometimes my mind was just wandering around crazily with negative thoughts, and I really hope that I can learn to control them properly." Again, it is normal to lose count. It is difficult because we are training our brains.

- I consider this kind of meditation a form of learning. You learn to focus your attention, then you forget—because your brain wanders off into some story in your head. Then you remember that you forgot to pay attention. And you go back to counting the breath, starting again at one. And back again. *Learn, forget, remember, learn, forget, remember.* This is brain training. Do this over and over until your timer goes off. Do not look at the timer, and try not to worry too much about time or what you will have for dinner. Don't even worry too much about your legs. They will likely tingle and go to sleep, but from what I have heard, they will not fall off. The point here is to stay as focused as you can on your breath—the one you are taking this moment, right now. You have to train the brain like a muscle. If you are just sitting there thinking, this is not meditation.

- Once your timer goes off, stretch out your legs in front of you. They may feel tingly because they have gone to sleep. Not to worry. They just weren't getting enough oxygen. Make sure you move them around until they are not tingly anymore. And then slowly stand up.

- We are not quite done with the mental training part. Now we are going to do the second type of meditation— *walking meditation.* First, you need to find a quiet hallway or room with enough space to walk in. Ideally, it is the same space you were sitting in.

Set your timer again, this time for ten minutes.

- As before, the mental training is done in complete silence! No phones or music or TV. Just you and your own thoughts. If you are with others, stand in a big circle facing in one direction about four feet apart from one another.

- Now put your arms behind your back with your right hand holding your index and middle finger together from your left hand. Let them rest loosely on your backside. In this part of the training, you want to have your eyes open so that you don't run into anything or anyone. Keep your eyes lightly focused on the ground— just about three feet in front of your feet. Again, don't be looking around. Stay present.

- Now take your attention and bring it into your feet. Feel your feet. Wiggle them around a bit. Feel your feet against the floor. Now start to walk. Take your right foot a step forward. Feel the weight of your right foot go from the heel to the ball of your foot to the toes, and then feel the weight of your body shift to the other side as your left foot comes forward. And mostly, walk very, very slowly. This is not a race. We are just using the feet to focus our attention. Try to keep your attention in your feet. When your attention starts to wander around, notice and bring it back into your feet.

- Slow walking meditation is very similar in concept to sitting meditation. The primary difference is that we direct our attention into our feet rather than on our breaths. When our brains start to wander around, because we forgot to pay attention, we remember that we forgot to pay attention and we go back into our feet. *Learn, forget, remember, learn, forget, remember.* This is brain training. Do this over and over again. When your timer goes off, you have finished the "mental" part of MAP Training.

PHYSICAL TRAINING:
THIRTY MINUTES OF AEROBIC EXERCISE

- Now we quickly switch to physical training. Put on your workout shoes and find a good place to get busy—and sweaty. This could be a gym or a room in your house, or you could go outside. Again, you want to do this part of the program immediately after the mental training. Try to decide what you are going to do and where you will do it before you begin the program. What kind of aerobic exercise works for you? What kind do you like, and what kind do you think you will actually do? Many people like to run—either on a treadmill or outside on the road. This is obviously the easiest to achieve. But you can also be inventive and engage in a spinning class or make up your own interval training or aerobic dance routine. In our in-person and online *MAP Train My Brain* classes, we do choreographed athletic movements to music—which are definitely aerobic, but also fun!

- The physical training must be effortful, and by effortful I mean aerobic. The only way to know for sure is to take your own heart rate. Put your two fingers (index and middle finger) on the right side of your neck and press in until you feel your pulse. Once you locate it, you will know you are feeling the blood pulsing through your carotid arteries, which take oxygen from your heart into your brain. Keep your fingers pressed on your neck and then locate a second hand on a clock or timer. Start the timer and count the number of beats in ten seconds. Then multiply that number by six. This number is the number of beats your heart generates each minute.

• Please take your own heart rate—don't use a heart rate monitor on a watch or app. It is good to become more aware of your own heart—which is achieved through a process called *interoception*. We do aerobic exercise during MAP Training to get more oxygen to our brains *while* learning who we are, and that includes knowing what our hearts are doing.

• Take your heart rate now, before you start to exercise. The number will likely be less than 100, which corresponds to 100 beats per minute. This is your resting heart rate.

Set your timer for thirty minutes.

• Start with a bit of a warm-up (five minutes at most) with some stretching and slow movements. Then start to really work it out! Whatever you are doing, you must stay safe and yet put in effort. If you are running, run hard. If you are spinning, spin hard. If you are dancing, keep your arms and legs up and moving. If you can, do lots of jumps and squats and kicks. These physical movements get the heart rate up. But if you have bad knees or a bad back, please don't jump or kick. You can still get the heart rate up by moving your arms up and down in the air. Swimming is also a good way to get the heart rate up without high-impact activity. When you have been exercising hard for about twenty minutes, measure your heart rate. This number should be greater than 100 and, for most of us, ideally greater than 120 beats each minute. The exact number depends on your age and your aerobic capacity. The younger you are, the faster your heart should beat when you exercise. (Again, if you have any health concerns, please check in with your physician beforehand and, if cleared to go, make sure to calculate your aerobic zone, as I described earlier.) If your heart rate is not in your aerobic zone, then next time you need to work harder, but again, while staying safe.

- Keep your body going hard for another five minutes and then start to slow down a bit. Use the last five minutes to slow your heart down. Return the focus of attention back to your breath and into your feet. For our in-person and online classes, I typically play a slower song and walk in a circle at the end, bringing attention back into the feet while maintaining some attention on the breath. Take a couple of deep breaths with your hands outstretched. Now just bring your hands together and bow. This is a good way to thank yourself for MAP Training your brain today.

- That's it—the end of one session. MAP Training takes only one hour. But the benefits last longer, much longer indeed!

10

Why We Should Train Our Brains

Make the most of yourself for that is all there is of you.

—Ralph Waldo Emerson

Now that I have described *how* to MAP Train your brain, let me tell you what it can do for you. Of course, I cannot guarantee that it will make you happy or find the love of your life. But I can guarantee that afterward, you will feel different, and most likely you will feel better. I have never heard of anyone who felt worse after MAP Training. One woman wrote:

I want to share my story of healing with you. About five years ago, at age forty-two, I began having flashbacks. I went into therapy, which has helped me immensely, and during some of our sessions, my therapist encouraged me to focus on my breathing. I am also a marathoner, and I had noticed running was helpful. I was just starting to meditate when I came across your research. My mind shouted, "I knew it!" I am excited to be practicing MAP Training myself, and want to teach it to other runners, many of whom say they are likewise stressed and depressed.

Another wrote:

Over the course of the past year I have been suffering from depression. I both meditate and work out, but I rarely do them together. When I heard about your study, I started doing them together, and I noticed an immediate difference. My depression has improved dramatically, and I have regained motivation.

But these are anecdotes. I am sure you would prefer real evidence. Let me start by telling you about a study we did some years ago to determine whether this program would help people with feelings of depression.

LESS DEPRESSION AND FEWER RUMINATIONS

Depression is a serious problem for millions of Americans and people all around the world. In fact, as many as one in five Americans suffer with depression in their lifetime.[1] Medications help some people, but for many, they don't work or don't work completely. Neuroscientists don't really know exactly what causes depression—possibly some genetics, but mostly it seems to be learned or circumstantial. As discussed in chapter 3, trauma can cause depression to emerge, and then once it settles in, it can be very difficult to dislodge. And just when you think it is gone, it comes back with a vengeance. We all have felt depressed at some point in our lives, and thankfully most of us eventually recover. But for some, the feelings linger on and on.

Why some people recover from depression and others do not is rather a mystery. However, as I have discussed, we know that people who are depressed tend to ruminate—and often. If we could find a way to lessen those thoughts, maybe we could lessen the grip depression can have on the brain. To test this hypothesis, my research team got together with exercise scientist Brandon Alderman to recruit people who were clinically depressed; they were diagnosed with major depressive disorder.[2] We also recruited a group of people who were not depressed and otherwise healthy. Everyone engaged in MAP Training twice a week for eight weeks. We measured their thoughts and feelings before they started and then measured them again afterward. Amazingly, depressive symptoms decreased by nearly 40 percent. This is as good a response or perhaps better than most therapies, including antidepressants. Even the "healthy" controls were less depressed. The results from this study have been highly cited and were featured in *The New York Times* and NBC News, among other media outlets. So yes, MAP Training does help with depression.

What about rumination? Yes, the participants said they were engaging in fewer ruminative thoughts. The "healthy" controls were ruminating less as well. Apparently, engaging in one hour of meditation and aerobic exercise twice a week was sufficient to change thoughts, especially those negative repetitive ones about ourselves and our past. And these changes in thought patterns were reflected in the brain. During one test, participants were instructed to focus their gaze on an arrow in the middle of the screen and identify its direction. Meanwhile, this middle arrow was surrounded by many others, all pointing in different directions, making it more difficult to focus on the target arrow. While participants completed this attention task, we placed dozens of electrodes all over their scalp to measure brain activity. As mentioned earlier, the people who were ruminating more produced less synchronized activity in response to the arrow than people who were not ruminating as much. However, after MAP Training, their brains produced more of this activity.[3] Essentially, more neurons were firing at the same time in response to the focus of attention—the arrow. In general, this is a good thing. You want more neurons to fire in synchrony (at the same time) when you are paying attention to what is happening now. It is how things get done in the brain, and according to our data, MAP Training helps the brain do that.

WHAT ABOUT TRAUMA? DOES MAP TRAINING HELP?

These positive effects of MAP Training on depression were encouraging. The combination of meditation and aerobic exercise helped people feel less depressed while decreasing their tendency to ruminate. But what about trauma? Can it help people recover from trauma? Can it help people focus less on traumas in their past? I had developed the *MAP Train My Brain* program for this purpose, but initially could not find a good way to test it. Then one day, I happened to meet Peg Wright, the founder and CEO of the Center for Great Expectations. Her mother was forced to give her up for adoption when she was just an infant, and as a result, Peg has devoted her life to helping women keep their children. She took on an even harder task—helping women who have been homeless become reunited with their children while overcoming the trauma they suffered along the way. At the center, the mothers live with their children 24/7 while receiving food, shelter, and trauma-informed care. Needless to say, it is an amazing place, and she is an amazing person.

Imagine being physically and sexually traumatized as a teenager and then finding yourself homeless and living in a tent city in Toms River, New Jersey. Imagine having two hungry children to feed and another on the way. Imagine what this kind of suffering does—day in and day out—to your brain, the human brain. Imagine what it does to the brains of your children. It's hard to imagine because most of us have not come close to experiencing this kind of trauma. And these women have been through so much of it. One woman pointed out a small piece of grass by the train station, where she would try to sleep at night, bundled up in an old coat. She would go to the station every morning asking commuters for money or food; most ignored her. It was therefore not a surprise to find that many of the women were experiencing symptoms related to trauma—depression, anxiety, addiction, and a lot of blame.

So, we got to work. We held the MAP Training program in the basement, which was not easy. There were often babies in the room and interruptions of all kinds. But we persevered, and after just eight weeks, the women were less depressed, less anxious, and were engaging in more loving interactions with their children. And their physical health improved dramatically.[4] We measured VO_2, which estimates how much oxygen is consumed by the body with each breath; theirs increased on average by 40 percent.

THE WHOLE IS GREATER THAN THE SUM OF THE PARTS

You might be saying to yourself: These results aren't that surprising; everyone knows that meditation is good for us, as is aerobic exercise. You might be wondering: what's so special about the combination? Are they better done together than alone? Could someone just do one activity and be better off? We tested this idea some years ago.[5] First, we recruited a large group of women, many of whom had experienced sexual violence and were experiencing symptoms related to trauma, including trauma-related thoughts and ruminations. The women were placed into one of four groups. One group engaged in MAP Training twice a week, which consisted of thirty minutes of meditation followed immediately by thirty minutes of aerobic exercise. The second group meditated twice a week. A third group exercised twice a week. Another group did neither activity. After six weeks, the women who did both activities together reported many fewer thoughts about their traumas—the so-called trauma-related cognitions that I have talked about. They also reported fewer ruminations—those nagging negative thoughts about ourselves. They even reported greater self-worth, meaning they felt better about who they are now than they did before they started the training. The women who only exercised or only meditated did benefit, but not as much as the women who did both activities together.

Therefore, doing both mental and physical training together, one after the other, is generally better than doing just one of them. As the saying goes, the whole is greater than the sum of the parts.

I was pleased with these results, but not surprised. When I do these two activities myself one after the other, I feel better than after doing just one of them. After aerobic exercise alone, I feel excited and energetic, but sometimes I have almost too much energy—right on the edge of anxious. After meditation on its own, I feel less anxious, but I don't necessarily have more energy. When I do both activities, one after the other, I feel almost euphoric and always more enthusiastic. Through the years, I have heard similar descriptions from others using MAP, most of them positive, and some of them profound. One woman told me that before the training, she had never thought about a thought before. She had always acted on her thoughts without really thinking about why she has them or how they are generated in the brain. Another told me that the training helped her with her sense of self. She said, "I used to be an addict, but I am not one now." Still another told me that she could actually read a book at night without getting distracted. I don't even need to look in their brains to know what they tell me is true.

"IT HELPS ME NOT THINK SO MUCH"

Participants often have questions about the program, one of which comes up more than any other: "Is there some reason you placed meditation before physical exercise in your protocol? Intuitively, I'd have thought that it would be better to exercise first since, once your heart rate normalized, you'd be invigorated for a concentration practice. Is there a neuroscience-based reason for doing meditation before aerobic exercise?"

Yes, it does feel easier to meditate after you exercise because you are relaxed and energized and generally "happier." This is exactly why we don't exercise first. The point of this program is to train the brain with effort, not to relax or enjoy some alone time, and so forth. To make it effortful, we must force ourselves to sit in silence with our own thoughts as they are when we first sat down, not as they are after a bout of aerobic exercise. The order is important. The point is to learn, forget, and remember to concentrate while at the same time having all those repetitive and annoying thoughts swimming around in our brains. After the mental training, we fill the brain with oxygen to consolidate the learning, at least theoretically. But I have another reason for insisting on the order: I wanted this program to be fast and efficient—one hour maximum. When people exercise aerobically in this way, they may want to change clothes, maybe get some water and check their phones. Before you know it, hours will have passed! So yes, order is important. Mental training comes before physical training. Always.

Again, I want to stress the fact that *MAP Train My Brain* is not a substitute for psychotherapy or medication but rather a behavioral health program that improves the fitness of the brain and body by exercising them both together, one after the other.[6] The benefits are many—but most consistent is the decrease in rumination. In fact, in one of our studies, people reported as much as a 25 percent decrease after training just once a week for six weeks! And not only that—the effects were also quite long-lasting—persisting up to six months later. Of course, not all of the participants were continuing to do the program on their own—some were, and some were not. But even so, it seems like they had learned along the way to ruminate less. These results are good on their own, but as you might recall, ruminative thoughts are tightly linked to other thoughts—such as trauma-related thoughts—and to other feelings, such as depression and anxiety. By decreasing these repetitive thoughts, one might presume that some of the other problems also fall away.

As discussed, women are much more likely to be diagnosed with stress-related mental illnesses, such as depression and anxiety and

PTSD. They are also more likely to ruminate than men are. And so again, by decreasing their prevalence, we might have a means for decreasing the prevalence of these illnesses in women. But to be clear, we all ruminate, regardless of whether we are a man or a woman or have been diagnosed or have even thought about getting a diagnosis. For example, we recently provided an eight-week MAP Training program to a large group of medical students, many of whom were struggling with their grueling schedules. In general, they were already mentally and physically quite healthy. Yet even for them, the MAP Training program reduced their inclination to ruminate.[7] And with this came a greater quality of life. My point here is that we all ruminate to some extent, and we can all learn to do it less often.

WHAT WE ARE LEARNING

When people ask me why I think MAP Training helps people, I have to imagine it is at least in part because they are learning a new way of experiencing their own thoughts—perhaps with less of a focus on memories from the past. In his essay "Baptism of Solitude," novelist Paul Bowles wrote about his time spent in the Sahara Desert: "In this wholly mineral landscape lighted by stars like flares; even memory disappears; nothing is left but your own breathing and the sound of your heart beating." I think of MAP Training in this way—as a way of learning to live alongside our thoughts in the absence of so many memories.

But what exactly are we learning while we sit and then walk in silence and then immediately exercise aerobically? Let's start with the most obvious form of learning. Throughout our lives, when we find ourselves in dangerous situations, we feel our hearts start to race. As discussed in chapter 3, we feel this response because our brains send a signal into the body to increase the beating of the heart, which then sends oxygen to our muscles, so that we can make a run for it, and to our brains so that we can make quick decisions and create memories. Because we are exposed to this response so often, our brains have learned to associate a beating heart with fear—and trauma. Even the thought of a past trauma can make our hearts race. But during MAP Training, our brains are learning something new. While we sit in silence trying to focus our attention on our breaths, thoughts come and go. Some of those thoughts are mundane and peaceful while others are full of stress and anxiety and some bring with them memories of trauma. But while thinking these thoughts, our hearts are beating slowly. As a result, we are learning that the thoughts and memories that come up during meditation—even traumatic ones—are not necessarily associated with a racing heart and therefore do not indicate pending danger. We are learning to *dissociate* traumatic thoughts from the feeling of a racing heart.

Then later, as we enter the exercise part of MAP Training, our hearts start to beat faster—and ultimately much faster. But we are not afraid, and we might even be having fun. Now our brains are learning another new association—that a racing heart is not necessarily associated with danger. Once again, we are learning to dissociate thoughts and feelings of fear from what is happening in our hearts. At first, the brain may not fully realize that it is learning these new associations, but with practice it will. Eventually, we learn to be less afraid of a pounding heart and more appreciative of a quiet brain. The ways we can put this new learning to use are infinite. Let me now tell you about one.

DISTINGUISHING OLD FROM NEW

The brains of people who experience trauma tend to generalize—meaning that they learn to be afraid about not only the actual events that were around during the trauma but other similar events or stimuli. Initially, this is an adaptive response, but over time it can begin to interfere with everyday life and can perpetuate some of the symptoms of PTSD. A recent study examined this process in people who had gone to the emergency room, mostly because of car accidents. Two months after the trauma, they were brought into a lab and trained to distinguish a fearful context from one that was safe while their brains were being scanned for activity in the hippocampus. The people who reported the fewest symptoms of PTSD had the most activity in their hippocampus. It's as if they were using neurons in this brain region— new and old ones together—to help them distinguish a threatening context from a safer one.[8]

Recall that the MAP Training program was inspired by the discovery of new neurons in the hippocampus, which increase in number after learning and aerobic exercise, at least in laboratory studies. But you might wonder what is so special about a new neuron, particularly if it just ends up connected to all the other old neurons in the brain. Neuroscientists don't know for sure, but I have a hypothesis. The hippocampus is designed for learning, especially learning about new and meaningful experiences that are happening right now in this moment. It is also used for encoding the context of this new experience—"the when and where." How does a cell know that an experience is new? If the neuron has been in the brain for years and years, the cell or group of cells would need to have a memory of everything that has ever happened before, which would take up a lot of "space," at least theorectically. But a new neuron has no experience. For it, everything is new, a feature that could come in handy when trying to figure out what is old from what is new.

So, how might the brain use new neurons to help us navigate our everyday lives? Maybe they help us distinguish old thoughts and memories from newer ones, which we do all the time without necessarily being aware.[9] Say you need a new printer, and you go to the local Best Buy to pick one up. You've been in this store before, but not this exact one. Your brain automatically retrieves the memory of your previous trips, and though you know this isn't the exact store you were in, you start moving toward the section where you've seen printers located previously—you are being guided by the old memory in a new store. You have generalized the memory of the store so you can use it in the new one. Or imagine once again that you are walking down a dark street alone at night and sense someone coming up close behind you. You become afraid because something bad happened to you before on a dark street. Your heart starts to race, and you consider making a run for it. But then you hear a voice. Your brain compares the voice you hear now to other voices you know and recognizes it as your neighbor's. Now you know that you do not need to be afraid. You relax and catch up with your neighbor.

Engaging in the *MAP Train My Brain* fitness program seems to enhance the ability to make these kinds of fine distinctions between the old and the new.[10] In one of our studies, participants were shown pictures of complex yet common objects (chairs, umbrellas, trees, etc.) and asked to categorize them as either "indoor" or "outdoor" with a button press. Then later, participants were given a surprise recognition test in which they were shown repetitions of previously viewed objects, objects similar but not identical to those in the prior set, and novel objects. And they had to say whether the objects were the same, similar, or new. This task is quite difficult to do, but after six weeks of MAP Training, once a week, participants were generally better at distinguishing the memory of the old image from the new one, even though the objects themselves were nearly identical. Like many human experiments, these kinds of comparisons may seem rather obtuse, but we can use them to generalize about situations that are meaningful. For example, one person told me how the training program helped her with one very specific memory. She was haunted by images from the day she followed some friends into a house, where she became infected with HIV. She used to play that day over and over in her brain, trying to imagine herself not going inside and instead just walking by. Now she walks down her street, the same street, and does not follow her old memory into the house. She chooses to experience this moment as new.

11

Living with Traumas:
Past, Present, and Future

The past doesn't really exist anymore . . . except in our brains.

—Professor Shors

I created the *MAP Train My Brain* fitness program because I wanted to offer a program that helps people lead better lives but does not require a lot of time or money or fancy equipment. I also wanted it to help all people, irrespective of gender, age, or ethnicity. According to our studies so far, it does. But I am a realist. This program is not perfect, and it will not solve all our problems. No one program or practice or therapy will. We have to learn new skills and then keep learning more and more new skills throughout our lifetimes to keep our brains fit. Then when traumas arise, and they will, we will be ready with the right skill for the right thought in the right moment.

To illustrate, let me tell the "Cliffhanger Story." A woman is calmly strolling through the woods, when she sees a tiger and starts to run for her life. She is running and running, and the tiger is getting closer and closer. She reaches the end of the trail and has no choice but to jump off a cliff. She jumps and is lucky enough to catch a vine on the way down. As she hangs on the vine, she is relieved—at least for the moment. However, when she looks up, she sees the tiger looking at her over the ledge of the cliff, and when she looks down, she sees a deadly fall. Then she sees two little mice peek their heads out of the side of the cliff. They start to chew on her vine. Her heart is racing even more than it was—she realizes that she is going to die. And then she sees a big strawberry growing right out from the vine. She grabs the strawberry and takes a bite. She says to herself, "What a delicious strawberry."

What do you think this story means? Most people think it means that you need to live in the moment and enjoy it, especially if it happens to be a good one. This is what I thought the story meant when I first heard it. This interpretation would certainly be consistent with the way many people think of "mindfulness"—as being mindful enough to appreciate this moment. But let's consider other interpretations. For example, perhaps the story really means that the woman hanging on the vine is now distracted by her thoughts of pleasure (the strawberry), and while being distracted, she is not looking for other potential routes of escape. Were she not so busy savoring the idea of the strawberry, she might have seen another branch weaving down the cliff and safely away from the tiger. Or perhaps she could have given the strawberry to the mice so they would quit gnawing on her vine.

My point here is not to endorse one interpretation of the story over another but rather to underscore the variety of responses we can make in each moment. What are we really doing in each moment? Are we prepared to make a decision, or are we lost in the past? Or are we worrying about the future? Are we wishing things could be different and therefore missing the opportunity to make wise choices—not necessarily the most fun but the wisest? The famous American psychologist William James said,

The great thing, then, in all education, is to make our nervous system our ally instead of our enemy. It is to fund and capitalize our acquisitions, and live at ease upon the interest of the fund. For this we must make automatic and habitual, as early as possible, as many useful actions as we can, and guard against the growing into ways that are likely to be disadvantageous to us, as we should guard against the plague. The more of the details of our daily life we can hand over to the effortless custody of automatism, the more our higher powers of mind will be set free for their own proper work.[1]

THE MENTAL SKILLS: NEW YET OLD

I don't know about you, but I feel completely inspired by these words from William James more than a century after he wrote them. So how can we put them to work? How can we make it a real and realized part of our lives? Let's consider a series of studies during which people were asked to manipulate their thoughts about memories in real time while their brains were being imaged.[2] In general, they were presented with various images and then later were asked either to linger with the memory of the image or try to suppress it. Or they were asked to replace the memory with another image or clear their minds of the image altogether. Each of the mental skills activated a different brain network, and some were easier to implement than others. Some skills were more disruptive to ongoing brain function than others. For example, trying to clear the memory was especially difficult to do and was more likely to interfere with ongoing mental processes. These are compelling findings, but the skills themselves are not new. In fact, you may have noticed similar skills—linger, replace, suppress, clear—are frequently engaged during various forms of trauma therapy, at least theoretically. A client undergoing prolonged exposure (PE) therapy may be asked to consciously linger with a negative memory in a safe environment until the associated fear begins to dissipate, whereas someone going through cognitive processing is encouraged to restructure their beliefs about what happened in the past. During the empty chair technique, the client speaking to the empty chair begins to "see" their memories of an experience from another perspective, similar but different. In general, the idea is to learn these new mental skills during therapy and then practice them in everyday life.

These kinds of mental skills are not new to psychology either. They have been practiced for thousands of years during meditation, which trains us to use our brains in a new way. While sitting in *open awareness,* people are asked to take in all thoughts and sensations without judgment, which is not easy to do. Or consider what happens during the *benefactor* practice. People are told to replace a negative thought with the image of someone they love. When I tried this, I found it useful to imagine my favorite uncle driving up to take me for a ride in his red convertible. And then there is *MAP Train My Brain.* For this program, I borrowed from a focused-attention practice, during which people are instructed to focus on the tiny space between the out-breath and the in-breath. When the mind wanders off, they are told to notice and replace the wandering thought with the tiny space.

But how do these mental skills work to change the brain? How can they help us feel less stress while ruminating less often on the past? We may never know the exact mechanisms but we can be sure of one thing: the brain is always changing and always learning and if given the chance, is ready to remember and use what it has learned. For example, when an autobiographical memory or negative thought about a memory arises, an emotional response may initially rise but then, over time, may lessen in intensity as the participant "learns" that the two events are not necessarily or always associated. An interruption of these learned associations with new learning may "simply" override the learned tendency to ruminate so often on the past. To be clear, I am not claiming that any one therapy or any one meditation practice engages one skill to the exclusion of others, and I am certainly not endorsing one approach over another. Quite the opposite. I recommend learning as many of these skills as you can. Who knows when you might need to use one of them?

TRAINING OUR BRAINS FOR THE FUTURE

The year 2020 was going to be promising. Then before I could put away my holiday decorations, I was waiting with a crowd of people in an emergency room, all of us struggling to breathe. After a chest x-ray, I was sent home with some prednisone (a type of cortisol) for the inflammation in my lungs. Once I got home, I turned on the news to hear story after story of people getting sick and dying from the coronavirus. I was convinced I had it but could not get tested. The more I thought about the virus, the more I couldn't breathe. *What if I end up on a ventilator? Worse yet, what if my son gets sick and ends up on a ventilator?*

One night around three in the morning, I woke up my son and made him take me back to the emergency room. By then, all the doctors and nurses were wearing hazmat gear, and my son was forced to leave the hospital. I started to imagine never seeing him again, but luckily my chest x-ray was clean, and I was sent home, again with prednisone. However, this time I remembered, *I have a skill that might help me.* And so I used it. I lay in bed counting the space between my breaths, slowly one after the other. When my mind would wander off into negative thoughts, I would replace the thought by directing my attention back onto the space. It was extremely difficult to do, but I did it, and I am convinced it helped get me through the worst of it.

The coronavirus pandemic is an example of an everyday trauma, if there ever was one. As many as one in five people who were hospitalized experienced symptoms of PTSD—intrusive thoughts, traumatic memories, and anxiety.[3] Some people even had breaks with reality. And nearly all of us felt the loss of human connection and the comfort of knowing what tomorrow would bring. I never could get tested, but when I eventually recovered from whatever I had, I wanted to help others learn some of the skills that helped me. My graduate student and I decided to focus our efforts on elementary and high school teachers, who were out of school for the summer and preparing to go back in the fall. We recruited about fifty teachers, who took an online MAP Training class once a week. After just six weeks, they were feeling much less stressed and having fewer feelings of anxiety. They also said they were ruminating less often.[4]

In general, their responses to MAP Training were similar to what we had observed with other populations—but they differed in one important way. The training program began *before* cases of coronavirus surged in the United States and well before teachers had to go back to school. During this time period, teachers had been asked to devise methods for protecting themselves and children in the classroom. Some even built Plexiglas barriers for their desks. Still others were being asked to completely redo their teaching materials for online instruction. We predicted that under these conditions, teachers were likely becoming more stressed and anxious as the new school year approached—and many were. In contrast, the teachers who were training their brains during this same time period did not get worse. Quite the opposite. They actually reported feeling better, much better. You might say they had become more resilient.

One of the participants, Jessica, had been diagnosed with an anxiety disorder years ago and had recovered. During the pandemic, she felt herself becoming more and more anxious. "I have been having a lot of trouble falling asleep. But I find that if I use your meditation technique, I can calm my mind enough to fall asleep. Thank you so much for that! In the coming weeks as the beginning of school approaches, I am sure I will be using it often." These results were even better than I had hoped! But to be clear, these are not *my* techniques, and I certainly didn't discover them. They are ancient. We just have to learn how and when to use them.

EXPECTING THE UNEXPECTED

I used to spend hour upon hour hunched over a microscope looking for traces of memories in slices of brain tissue. Yet as soon as I left my lab, I would immediately start ruminating, going over old memories and completely ignoring everything I knew as a scientist about the brain. It is really hard to accept the fact that our memories are located in our brains and generated by ions crossing neuronal membranes. And that this process is essentially how we create the stories of our lives. So, if we really want to create better stories, we have to embrace the facts and change our brains. And I don't mean this in some kind of catchphrase pop psychology way, but rather in the sense that we should more fully appreciate what the brain actually does and how it works. But to do this, we have to go beyond getting more sleep or counting our breaths or doing jumping jacks. It is not enough. We have to treat our brains better than we treat our hair or our biceps. We have to recognize the kinds of thoughts and feelings we are having day to day and then use that information to train our brains with intention, with passion, and with compassion.

I started this book with one of my mom's favorite sayings: "Everyone has a story." So, what are your stories? What are the stories you tell when you feel like you can't handle what's going on in your life? And can you see those stories changing over time? I hope that by learning a bit about how the brain goes about creating those stories and then replaying them, you have gained some insight into the many processes at work as we travel from thoughts to memories to feelings back to thoughts, memories, feelings, back and forth again and again. The goal isn't to stop thinking or stop feeling or even stop remembering. It's just to see these processes evolve and accept them for what they are. As we start to see them for what they are, we can let some of the stories go, while embracing others.

My mother had another favorite saying: "Hope for the best, but expect the worst." My family never resonated with this one as much as I do; they find it too depressing. But I consider it profound. We have to be realistic and learn to expect the unexpected. With that, let me share with you one last story about Maria, a young woman who told me about a traumatic experience she had as a child. Her family was sleeping soundly in their apartment, which was on the tenth floor of an old high-rise, when her parents ran frantically into her room to wake her up.

> I felt the building start to shake, and then my dad yelled out, "Let's go!" We all went running out the exit door and down flight after flight of stairs. I was only focused on my survival. My legs felt weak, and I could feel my heart racing. Eventually, we reached the bottom floor and went out the back entrance to see a huge sink-hole where our parking garage used to be. The only thing I could think of were my pet hermit crabs that I left in the apartment. But my parents wouldn't let me go back and get them. Once I settled down, I realized why. This event showed me that our lives can end quickly and without warning. It taught me to appreciate my family and friends— and pets. Luckily, nobody was seriously hurt, and my hermit crabs survived. I will never forget this day and feel very lucky because it could have turned out a lot worse.

We have to be realistic about what our lives will bring. All of it won't be good or fun or even neutral. Bad things are going to happen, traumas will continue to occur; we will lose people we love, our relationships will not work out, and of course we could experience another pandemic. As much as we try not to, we will still ruminate to some extent on the memories of these new traumas. But we should, meanwhile, reflect more on the good stories, revisiting the best times and loves of our lives, our accomplishments and whatever we have done along the way to fulfill our dreams. And mostly, we must prepare our brains for the stories we are creating now and the ones to come.

To turn all this into reality, we need to commit ourselves to learning new skills, both mental and physical, each and every day. It will not be easy. We will have to put in some effort. But these are desirable difficulties. They keep our brains stimulated and engaged in life while teaching them how to distinguish what is old from what is new, what is dangerous from what is safe. Engaged in this way while seeing things as they are, we will be ready for what comes, good and bad. Our brains will be ready because we are fit for life.

ACKNOWLEDGMENTS

As soon as I learned how to read, I wanted to write a book. And so, first and foremost, I thank my mother, who filled my brain with a love of stories and all things books. As for becoming a scientist, I credit my dad and brother Clay, who taught me to find facts wherever I could, especially if they answer questions about how things actually work in the world. And for learning a lust for life, I remember my sister Marcia. I'm grateful for Louis Matzel, who taught me more about learning than anyone and taught me even more about love, as we both watched our son, Evan Shors Matzel, become the man he is today. Thank you to Richard "Dick" Thompson for challenging me to think deeply about mechanisms of memory in the brain—and thanks to both him and his wife, Judith Thompson, for taking me into their lab and their lives as they would a daughter. Thank you to Stephanie White, my longest-living best friend with whom many stories continue to be made and then rehearsed—over and over. I wish to acknowledge Myong Ahn-Sunim, who taught me that doing meditation is better than talking about it, and two beams of light that went out all too soon: Rick Wilson, who showed me what it really means to dance aerobically, and Eric Arauz, who shared his lived experience, traumatic as it was, to help others with theirs.

As I moved from the laboratory bench out into the world, I was beyond lucky to have at my side Drs. Emma Millon and Michelle Chang, who conducted our human studies with compassion, understanding, and attention to detail. No one, and I mean no one, could have done it better. Along the way, I have had many dedicated students, scholars, and collaborators, most of whom are now doctors and scientists out on their own, making the world a better place— Megan Anderson, Debra Bangasser, Anna Beylin, Dani Curlik, Christina Dalla, Docia Demmin, Gina DiFeo, Erik Dryver, Demetrius Durham, Jacqueline Falduto, Georgia Hodes, Paul Lavadera, Benedetta Leuner, Caroline Lewczyk (now Boxmeyer), Lisa Maeng, Sabrina Mendolia-Loffredo, George Miesegaes, Miriam Nokia, Jane Pickett, Jessica Santollo, Joel Selcher, Rick Servatius, Helene Sisti, Krishna Tobon, Jaylyn Waddell, Gwendolyn Wood, Mingrui Zhao, and others I might have forgotten to name.

It has been and remains my privilege to be a member of the faculty at Rutgers University and to have been awarded scientific funding from the National Institutes of Mental Health at the National Institutes of Health, the National Science Foundation, the National Aeronautics and Space Administration, the National Alliance for Research on Schizophrenia and Depression through the Brain and Behavior Research Foundation, and the Brain Health Institute at Rutgers University. Thank you, Elizabeth Gould, for letting me in on one of the biggest discoveries of the century. I am indebted to Brandon Alderman and his graduate students Ryan Olson and C. J. Brush for getting serious about the neuroscience of MAP Training. Thank you, Peg Wright and Beata Zita, for welcoming me and my team into their centers and hearts, and for introducing me to the amazing people who participated in our studies. I wish to acknowledge Lauren Linscott and Rebecca Vazquez at Rutgers' Office of Violence Prevention and Victim Assistance for all you and your group do for survivors. For grit and determination, I look to Roberta Diaz Britton and Catherine Woolley, science superstars whom I am proud to call close friends.

For this book, I am indebted to Flatiron Books and Macmillan for trusting me to get it done during a global pandemic. A special thank-you to Julia Coopersmith for helping me translate my ideas into a proposal and showing me how to write for a general audience, and to my first editor, Sarah Murray, who set me off, and my last editor, Bryn Clark, who got me to the finish line. From there, my gratitude extends to colleagues and friends who took time out of their already busy schedules to read and comment on sections along the way: Dani Curlik, Docia Demmin, Samantha Farris, Megan Giles, Evan Kleinman, Louis Matzel, Emma Millon, and Daniel Ogilvie. Thank you, Esther Bennett, for your honesty, Emilie Rissman for your support, and Edward Selby for suggesting years ago that I look into ruminations. I have to wonder what the book would even be about were it not for this suggestion.

It's not easy to write a book about trauma. And so finally, I want to recognize Jaidree Braddix, the best literary agent ever, and Celeste Fine, a pro among pros. Both encouraged me to write a book that would matter to others while enjoying the process. I hope that I have and, thanks to them, I am.

NOTES

PROLOGUE

1. R. A. Lanius, J. W. Hopper, and R. S. Menon, "Individual Differences in a Husband and Wife Who Developed PTSD After a Motor Vehicle Accident: A Functional MRI Case Study," *American Journal of Psychiatry* 160, no. 4 (2003): 667–69.

1. LIFE'S TRAUMAS—BOTH LARGE AND SMALL

1. R. C. Kessler, S. Aguilar-Gaxiola, J. Alonso, C. Benjet, E. J. Bromet, G. Cardoso, L. Degenhardt, G. de Girolamo, R. V. Dinolova, F. Ferry, S. Florescu, O. Gureje, J. M. Haro, Y. Huang, E. G. Karam, N. Kawakami, S. Lee, J. P. Lepine, D. Levinson, F. Navarro-Mateu, B. E. Pennell, M. Piazza, J. Posada-Villa, K. M. Scott, D. J. Stein, M. T. Have, Y. Torres, M. C. Viana, M. V. Petukhova, N. A. Sampson, A. M. Zaslavsky, and K. C. Koenen, "Trauma and PTSD in the WHO World Mental Health Surveys," *European Journal of Psychotraumatology* 8 (2017).
2. C. Benjet, E. Bromet, E. G. Karam, R. C. Kessler, K. A. McLaughlin, A. M. Ruscio, V. Shahly, D. J. Stein, M. Petukhova, E. Hill, J. Alonso, L. Atwoli, B. Bunting, R. Bruffaerts, J. M. Caldas-de-Almeida, G. de Girolamo, S. Florescu, O. Gureje, Y. Huang, J. P. Lepine, N. Kawakami, V. Kovess-Masfety, M. E. Medina-Mora, F. Navarro-Mateu, M. Piazza, J. Posada-Villa, K. M. Scott, A. Shalev, T. Slade, M. ten Have, Y. Torres, M. C. Viana, Z. Zarkov, and K. C. Koenen, "The Epidemiology of Traumatic Event Exposure Worldwide: Results from the World Mental Health Survey Consortium," *Psychological Medicine* 46, no. 2 (2016): 327–43.

3. National Comorbidity Survey, "NCS-R Lifetime Prevalence Estimates," https://www.hcp.med.harvard.edu/ncs/ftpdir/NCS-R_Lifetime_Prevalence _Estimates.pdf.

4. General resources on the topic: Hans Selye, *The Stress of Life* (New York: McGraw-Hill, 1956); Judith Herman, *Trauma and Recovery* (New York: Basic Books, 1992); Richard J. McNally, *Remembering Trauma* (Cambridge, MA: Belknap Press, 2003); Bessel Van der Kolk, *The Body Keeps the Score: Brain, Mind, and Body in the Healing of Trauma* (New York: Penguin, 2014); Robert M. Sapolsky, *Why Zebras Don't Get Ulcers* (New York: W. H. Freeman, 1994).

5. National Center for PTSD, "How Common Is PTSD in Women?," https:// www.ptsd.va.gov/understand/common/common_women.asp.

6. Eric C. Arauz, *An American's Resurrection: My Pilgrimage from Child Abuse and Mental Illness to Salvation* (Newark, NJ: Treehouse Publishing, 2012).

7. World Health Organization, "Violence Against Women," https://www.who.int /health-topics/violence-against-women#tab=tab_1.

8. G. Ayano, M. Solomon, L. Tsegay, K. Yohannes, and M. Abraha, "A Systematic Review and Meta-analysis of the Prevalence of Post-traumatic Stress Disorder Among Homeless People," *Psychiatric Quarterly* 91, no. 4 (2020): 949–63.

2. HOW STRESS AND TRAUMA CHANGE OUR LIVES

1. American Psychiatric Association, *Diagnostic and Statistical Manual of Mental Disorders,* 5th ed. (Washington, D.C.: APA Publishing, 2013); T. M. Keane, A. D. Marshall, and C. T. Taft, "Posttraumatic Stress Disorder: Etiology, Epidemiology, and Treatment Outcome," *Annual Review of Clinical Psychology* 2 (2006): 161–97; Julian D. Ford, *Posttraumatic Stress Disorder, Scientific and Professional Dimensions,* 2nd ed. (Cambridge, MA: Academic Press, 2015).

2. E. M. Millon, H. Y. M. Chang, and T. J. Shors, "Stressful Life Memories Relate to Ruminative Thoughts in Women with Sexual Violence History, Irrespective of PTSD," *Frontiers in Psychiatry* 5, no. 9 (2018): 311.

3. Joseph LeDoux, *Anxious* (New York: Oneworld Publications, 2015); Jeffrey Gray and Neil McNaughton, *The Neuropsychology of Anxiety: An Enquiry into the Functions of the Septo-Hippocampal System,* 2nd ed., Oxford Psychology Series 33 (Oxford: Oxford University Press, 1982).

4. R. M. Nesse, "What Good Is Feeling Bad? The Evolutionary Benefits of Psychic Pain," *The Sciences* (November–December 1991): 30–37.

5. D. S. Hasin, A. L. Sarvet, J. L. Meyers, T. D. Saha, W. J. Ruan, M. Stohl, and B. F. Grant, "Epidemiology of Adult *DSM-5* Major Depressive Disorder and Its Specifiers in the United States," *JAMA Psychiatry* 75, no. 4 (2018): 336–46; K. Smith, "Mental Health: A World of Depression," *Nature* 515, no. 7526

(2014): 181; G. S. Malhi and J. J. Mann, "Depression," *Lancet* 392, no. 10161 (2018): 2299–312.

6. Andrew Solomon, *The Noonday Demon: An Atlas of Depression* (New York: Scribner, 2001), 443.

7. J. S. Buyukdura, S. M. McClintock, and P. E. Croarkin, "Psychomotor Retardation in Depression: Biological Underpinnings, Measurement, and Treatment," *Progress in Neuro-Psychopharmacology and Biological Psychiatry* 35, no. 2 (2011): 395–409.

8. P. D. Rozée and G. B. Van Boemel, "The Psychological Effects of War Trauma and Abuse on Older Cambodian Refugee Women," *Women and Therapy* 8, no. 4 (1990): 23–50; G. B. Van Boemel and P. D. Rozée, "Treatment for Psychosomatic Blindness Among Cambodian Refugee Women," *Women and Therapy* 13, no. 3 (1992): 239–66; S. Mattson, "Mental Health of Southeast Asian Refugee Women: An Overview," *Health Care for Women International* 14, no. 2 (1993): 155–65.

9. R. J. Loewenstein, "Dissociation Debates: Everything You Know Is Wrong," *Dialogues in Clinical Neuroscience* 20, no. 3 (2018): 229–42; C. J. Dalenberg, B. L. Brand, D. H. Gleaves, M. J. Dorahy, R. J. Loewenstein, E. Cardeña, P. A. Frewen, E. B. Carlson, and D. Spiegel, "Evaluation of the Evidence for the Trauma and Fantasy Models of Dissociation," *Psychological Bulletin* 138, no. 3 (2012): 550–88; G. A. Boysen and A. VanBergen, "A Review of Published Research on Adult Dissociative Identity Disorder: 2000–2010," *Journal of Nervous and Mental Disease* 201, no. 1 (2013): 5–11.

10. G. Weniger, C. Lange, U. Sachsse, and E. Irle, "Amygdala and Hippocampal Volumes and Cognition in Adult Survivors of Childhood Abuse with Dissociative Disorders," *Acta Psychiatrica Scandinavica* 118, no. 4 (2008): 281–90.

11. E. Vermetten, C. Schmahl, S. Lindner, R. J. Loewenstein, and J. D. Bremner, "Hippocampal and Amygdalar Volumes in Dissociative Identity Disorder," *American Journal of Psychiatry* 163, no. 4 (2006): 630–36; S. Chalavi, E. M. Vissia, M. E. Giesen, E. R. S. Nijenhuis, N. Draijer, G. J. Barker, D. J. Veltman, and A. A. T. S. Reinders, "Similar Cortical but Not Subcortical Gray Matter Abnormalities in Women with Posttraumatic Stress Disorder with Versus Without Dissociative Identity Disorder," *Psychiatry Research* 231, no. 3 (2015): 308–19; David Blihar, A. Crisafio, E. Delgado, M. Buryak, M. Gonzalez, and R. Waechter, "A Meta-analysis of Hippocampal and Amygdala Volumes in Patients Diagnosed with Dissociative Identity Disorder," *Journal of Trauma and Dissociation* 22, no. 3 (2021): 365–77.

3. THE TWO FORMS OF EVERYDAY TRAUMA

1. R. Mewes, H. Reich, N. Skoluda, F. Seele, and U. M. Nater, "Elevated Hair Cortisol Concentrations in Recently Fled Asylum Seekers in Comparison to

Permanently Settled Immigrants and Non-immigrants," *Translational Psychiatry* 7, no. 3 (2017): e1051.

2. Ibid.

3. B. McEwen, J. M. Weiss, and L. S. Schwartz, "Selective Retention of Corticosterone by Limbic Structures in Rat Brain," *Nature* 220 (1968): 911–12.

4. C. S. Woolley, E. Gould, and B. S. McEwan, "Exposure to Excess Glucocorticoids Alters Dendritic Morphology of Adult Hippocampal Pyramidal Neurons," *Brain Research* 531, nos. 1–2 (1990): 225–31.

5. A. V. Beylin and T. J. Shors, "Glucocorticoids Are Necessary for Enhancing the Acquisition of Associative Memories After Acute Stressful Experience," *Hormones and Behavior* 43, no. 1 (2003): 124–31.

6. B. Leuner and T. J. Shors. "New Spines, New Memories," *Molecular Neurobiology* 29, no. 2 (2004): 117–30.

7. J. B. Echouffo-Tcheugui, S. C. Conner, J. J. Himali, P. Maillard, C. S. DeCarli, A. S. Beiser, R. S. Vasan, and S. Seshadri, "Circulating Cortisol and Cognitive and Structural Brain Measures: The Framingham Heart Study," *Neurology* 91, no. 21 (2018): e1961–e1970.

8. J. M. Andreano and L. Cahill, "Glucocorticoid Release and Memory Consolidation in Men and Women," *Psychological Science* 17, no. 6 (2006): 466–70.

9. S. Schumacher, H. Niemeyer, S. Engel, J. C. Cwik, S. Laufer, H. Klusmann, and C. Knaevelsrud, "HPA Axis Regulation in Posttraumatic Stress Disorder: A Meta-Analysis Focusing on Potential Moderators," *Neuroscience and Biobehavioral Reviews* 100 (2019): 35–57; M. C. Morris, N. Hellman, J. L. Abelson, and U. Rao, "Cortisol, Heart Rate, and Blood Pressure as Early Markers of PTSD Risk: A Systematic Review and Meta-Analysis," *Clinical Psychology Review* 49 (2016): 79–91; R. M. Sapolsky, "Stress and the Brain: Individual Variability and the Inverted-U," *Nature Neuroscience* 18, no. 10 (2015): 1344–46.

10. S. Levine, "Maternal and Environmental Influences on the Adrenocortical Response to Stress in Weanling Rats," *Science* 156, no. 3772 (1967): 258–60; S. Levine, "Influence of Psychological Variables on the Activity of the Hypothalamic-Pituitary-Adrenal Axis," *European Journal of Pharmacology* 405 (2000): 149–60.

11. F. A. Champagne and M. J. Meaney, "Like Mother, Like Daughter: Evidence for Non-genomic Transmission of Parental Behavior and Stress Responsivity," *Progress in Brain Research* 133 (2001): 287–302; M. J. Meaney, "Maternal Care, Gene Expression, and the Transmission of Individual Differences in Stress Reactivity Across Generations," *Annual Review of Neuroscience* 24 (2001): 1161–92; F. A. Champagne, "Maternal Imprints and the Origins of Variation," *Hormones and Behavior* 60, no. 1 (2011): 4–11.

12. J. M. Goldstein, J. E. Cohen, K. Mareckova, L. Holsen, S. Whitfield-Gabrieli, S. E. Gilman, S. L. Buka, and M. Hornig, "Impact of Prenatal Maternal

Cytokine Exposure on Sex Differences in Brain Circuitry Regulating Stress in Offspring 45 Years Later," *Proceedings of the National Academy of Sciences* 118, no. 15 (2021): e2014464118.

13. B. D. Kelly, "The Great Irish Famine (1845–52) and the Irish Asylum System: Remembering, Forgetting, and Remembering Again," *Irish Journal of Medical Science* 188, no. 3 (2019): 953–58; Oonagh Walsh, *Insanity, Power and Politics in Nineteenth Century Ireland: The Connaught District Lunatic Asylum* (London: Manchester University Press, 2013).

14. R. Yehuda, S. M. Engel, S. R. Brand, J. Seckl, S. M. Marcus, and G. S. Berkowitz, "Transgenerational Effects of Posttraumatic Stress Disorder in Babies of Mothers Exposed to the World Trade Center Attacks During Pregnancy," *Journal of Clinical Endocrinology and Metabolism* 90, no. 7 (2005): 4115–18.

15. R. Yehuda and A. Lehrner, "Intergenerational Transmission of Trauma Effects: Putative Role of Epigenetic Mechanisms," *World Psychiatry* 17, no. 3 (2018): 243–57; A. Jawaid, K. L. Jehle, and I. M. Mansuy, "Impact of Parental Exposure on Offspring Health in Humans," *Trends in Genetics* 37, no. 4 (2021): 373–88.

16. P. E. Gold, "Regulation of Memory—From the Adrenal Medulla to Liver to Astrocytes to Neurons," *Brain Research Bulletin* 105 (2014): 25–35.

17. Walter B. Cannon, *The Wisdom of the Body* (New York: W. W. Norton, 1932); Steven W. Porges, *The Polyvagal Theory: Neurophysiological Foundations of Emotions, Attachment, Communication, and Self-Regulation* (New York: W. W. Norton, 2011).

4. RUMINATIONS: THOUGHTS THAT GET STUCK IN OUR BRAINS

1. E. B. Foa, A. Ehlers, D. M. Clark, D. F. Tolin, and S. M. Orsillo, "The Posttraumatic Cognitions Inventory (PTCI): Development and Validation," *Psychological Assessment* 11, no. 3 (1999): 303–14.

2. Edward Selby, personal communication with the author, June 29, 2021.

3. W. Treynor, R. Gonzalez, and S. Nolen-Hoeksema, "Rumination Reconsidered: A Psychometric Analysis," *Cognitive Therapy and Research* 27, no. 3 (2003): 247–59; S. Lyubomirsky, K. Layous, J. Chancellor, and S. K. Nelson, "Thinking About Rumination: The Scholarly Contributions and Intellectual Legacy of Susan Nolen-Hoeksema," *Annual Review of Clinical Psychology* 11 (2015): 1–22.

4. B. L. Alderman, R. L. Olson, M. E. Bates, Edward A. Selby, J. F. Buckman, C. J. Brush, E. A. Panza, A. Kranzler, D. Eddie, and T. J. Shors, "Rumination in Major Depressive Disorder Is Associated with Impaired Neural Activation During Conflict Monitoring," *Frontiers in Human Neuroscience* 9 (2015): 269; E. M. Millon, H. Y. M. Chang, and T. J. Shors, "Stressful

Life Memories Relate to Ruminative Thoughts in Women with Sexual Violence History, Irrespective of PTSD," *Frontiers in Psychiatry* 9 (2018): 311; T. J. Shors, E. M. Millon, H. Y. M. Chang, R. L. Olson, and B. L. Alderman, "Do Sex Differences in Rumination Explain Sex Differences in Depression?," *Journal of Neuroscience Research* 95, nos. 1–2 (2017): 711–18; P. Lavadera, E. M. Millon, and T. J. Shors, "*MAP Train My Brain*: Meditation Combined with Aerobic Exercise Reduces Stress and Rumination While Enhancing Quality of Life in Medical Students," *Journal of Alternative and Complementary Medicine* 26, no. 5 (2020): 418–23; E. M. Millon and T. J. Shors, "How Mental Health Relates to Everyday Stress, Rumination, Trauma and Interoception in Women Living with HIV: A Factor Analytic Study," *Learning and Motivation* 73 (2021).

5. S. Sütterlin, M. C. S. Paap, S. Babic, A. Kubler, and C. Vogele, "Rumination and Age: Some Things Get Better," *Journal of Aging Research* 2012 (2012): 267–327.

6. D. C. Blanchard, G. Griebel, R. Pobbe, and R. J. Blanchard, "Risk Assessment as an Evolved Threat Detection and Analysis Process," *Neuroscience and Biobehavioral Reviews* 35, no. 4 (2011): 991–98; D. C. Blanchard, "Translating Dynamic Defense Patterns from Rodents to People," *Neuroscience and Biobehavioral Reviews* 76, pt. A (2017): 22–28.

7. E. M. Millon et al., "Stressful Life Memories."

8. A. Whitmer and I. H. Gotlib, "Brooding and Reflection Reconsidered: A Factor Analytic Examination of Rumination in Currently Depressed, Formerly Depressed, and Never Depressed Individuals," *Cognitive Therapy and Research* 35, no. 2 (2011): 99–107.

9. C. P. Pugach, A. A. Campbell, and B. E. Wisco, "Emotion Regulation in Post-traumatic Stress Disorder (PTSD): Rumination Accounts for the Association Between Emotion Regulation Difficulties and PTSD Severity," *Journal of Clinical Psychology* 76, no. 3 (2020): 508–25.

10. C. Donaldson, D. Lam, and A. Mathews, "Rumination and Attention in Major Depression," *Behaviour Research and Therapy* 45, no. 11 (2007): 2664–78.

11. A. Curci, T. Lanciano, E. Soleti, and B. Rimé. "Negative Emotional Experiences Arouse Rumination and Affect Working Memory Capacity," *Emotion* 13, no. 5 (2013): 867–80.

12. B. L. Alderman et al., "Rumination in Major Depressive Disorder."

13. W. E. Mehling, M. Acree, A. Stewart, J. Silas, and A. Jones, "The Multidimensional Assessment of Interoceptive Awareness, Version 2 (MAIA-2)," *PLoS One* 13, no. 12 (2018): e0208034.

14. E. M. Millon and T. J. Shors, "How Mental Health Relates."

15. General references on topic: L. Nadel and O. Hardt, "Update on Memory Systems and Processes," *Neuropsychopharmacology* 36, no. 1 (2011): 251–73; J. L. C. Lee, K. Nader, and D. Schiller, "An Update on Memory Reconsolidation

Updating," *Trends in Cognitive Sciences* 21, no. 7 (2017): 531–45; E. A. Phelps and S. G. Hofmann, "Memory Editing from Science Fiction to Clinical Practice," *Nature* 572, no. 7767 (2019): 43–50.

5. THE BRAIN IS ALWAYS LEARNING

1. B. Leuner, J. Falduto, and T. J. Shors, "Associative Memory Formation Increases the Observation of Dendritic Spines in the Hippocampus," *Journal of Neuroscience* 23, no. 2 (2003): 659–65.
2. Donald Hebb, *Organization of Behavior: A Neuropsychological Theory* (New York: Wiley, 1949).
3. T. J. Shors and L. D. Matzel, "Long-Term Potentiation: What's Learning Got to Do with It?," *Behavioral and Brain Sciences* 20, no. 4 (1997): 597–614, discussion 614–55.
4. J. E. Dunsmoor, V. P. Murty, L. Davachi, and E. A. Phelps, "Emotional Learning Selectively and Retroactively Strengthens Memories for Related Events," *Nature* 520, no. 7547 (2015): 345–48.
5. R. F. Thompson, "In Search of Memory Traces," *Annual Review of Psychology* 56 (2005): 1–23; D. A. McCormick, G. A. Clark, D. G. Lavond, and R. F. Thompson, "Initial Localization of the Memory Trace for a Basic Form of Learning," *Proceedings of the National Academy of Sciences of the United States of America* 79, no. 8 (1982): 2731–35.
6. E. M. Millon, H. Y. M. Chang, and T. J. Shors, "Stressful Life Memories Relate to Ruminative Thoughts in Women with Sexual Violence History, Irrespective of PTSD," *Frontiers in Psychiatry* 9 (2018): 311.
7. Ibid.
8. J. S. Feinstein, M. C. Duff, and D. Tranel, "Sustained Experience of Emotion After Loss of Memory in Patients with Amnesia," *Proceedings of the National Academy of Sciences of the United States of America* 107, no. 17 (2010): 7674–79.
9. General references on topic: D. H. Zald, "The Human Amygdala and the Emotional Evaluation of Sensory Stimuli," *Brain Research Reviews* 41, no. 1 (2003): 88–123; R. L. Ressler and S. Maren, "Synaptic Encoding of Fear Memories in the Amygdala," *Current Opinion in Neurobiology* 54 (2019): 54–59; M. Davis, "The Role of the Amygdala in Fear and Anxiety," *Annual Review of Neuroscience* 15 (1992): 353–75; M. Zelikowsky, S. Hersman, M. K. Chawla, C. A. Barnes, and M. S. Fanselow, "Neuronal Ensembles in Amygdala, Hippocampus, and Prefrontal Cortex Track Differential Components of Contextual Fear," *Journal of Neuroscience* 34, no. 25 (2014): 8462–66.
10. C. S. Inman, K. R. Bijanki, D. I. Bass, R. E. Gross, S. Hamann, and J. T. Willie, "Human Amygdala Stimulation Effects on Emotion Physiology and Emotional Experience," *Neuropsychologia* 145 (2020): 106722.

11. J. L. McGaugh, "Time-Dependent Processes in Memory Storage," *Science* 153, no. 3742 (1966): 1351–58.

12. J. L. McGaugh, "Making Lasting Memories: Remembering the Significant," *Proceedings of the National Academy of Sciences of the United States of America* 110, suppl. 2 (2013): 10402–07.

13. C. S. Inman, J. R. Manns, K. R. Bijanki, D. I. Bass, S. Hamann, D. L. Drane, R. E. Fasano, C. K. Kovach, R. E. Gross, and J. T. Willie, "Direct Electrical Stimulation of the Amygdala Enhances Declarative Memory in Humans," *Proceedings of the National Academy of Sciences of the United States of America* 115, no. 1 (2018): 98–103.

14. A. Bechara, D. Tranel, H. Damasio, R. Adolphs, C. Rockland, and A. R. Damasio, "Double Dissociation of Conditioning and Declarative Knowledge Relative to the Amygdala and Hippocampus in Humans," *Science* 269, no. 5227 (1995): 1115–18.

15. E. M. Millon and T. J. Shors, "How Mental Health Relates to Everyday Stress, Rumination, Trauma and Interoception in Women Living with HIV: A Factor Analytic Study," *Learning and Motivation* 73 (2021): 101680; Millon et al., "Stressful Life Memories."

16. H. X. Zhou, X. Chen, Y. Q. Shen, L. Li, N. X. Chen, Z. C. Zhu, F. X. Castellanos, and C. G. Yan, "Rumination and the Default Mode Network: Meta-analysis of Brain Imaging Studies and Implications for Depression," *NeuroImage* 206 (2020): 116287.

17. T. J. Akiki, C. L. Averill, K. M. Wrocklage, J. C. Scott, L. A. Averill, B. Schweinsburg, A. Alexander-Bloch, B. Martini, S. M. Southwick, J. H. Krystal, and C. G. Abdallah, "Default Mode Network Abnormalities in Posttraumatic Stress Disorder: A Novel Network-Restricted Topology Approach," *NeuroImage* 176 (2018): 489–98.

18. M. I. Nash, C. B. Hodges, N. M. Muncy, and C. B. Kirwan, "Pattern Separation Beyond the Hippocampus: A High-Resolution Whole-Brain Investigation of Mnemonic Discrimination in Healthy Adults," *Hippocampus* 31, no. 4 (2021): 408–21.

19. Peace Pilgrim, *Peace Pilgrim: Her Life and Work in Her Own Words* (Santa Fe: Ocean Tree Books, 1992), 15.

6. WOMEN AND THEIR CHANGING BRAINS

1. T. J. Shors, "A Trip Down Memory Lane About Sex Differences in the Brain," *Philosophical Transactions of the Royal Society, London* 371, no. 1688 (2016): 20150124.

2. J. A. Clayton and F. S. Collins, "Policy: NIH to Balance Sex in Cell and Animal Studies," *Nature* 509, no. 7500 (2014): 282–83.

3. D. A. Bangasser and T. J. Shors, "Critical Brain Circuits at the Intersection Between Stress and Learning," *Neuroscience and Biobehavioral Reviews* 34, no. 8 (2010): 1223–33.

4. R. C. Kessler, S. Aguilar-Gaxiola, J. Alonso, C. Benjet, E. J. Bromet, G. Cardoso, L. Degenhardt, G. de Girolamo, R. V. Dinolova, F. Ferry, S. Florescu, O. Gureje, J. M. Haro, Y. Huang, E. G. Karam, N. Kawakami, S. Lee, J. P. Lepine, D. Levinson, F. Navarro-Mateu, B. E. Pennell, M. Piazza, J. Posada-Villa, K. M. Scott, D. J. Stein, M. T. Have, Y. Torres, M. C. Viana, M. V. Petukhova, N. A. Sampson, A. M. Zaslavsky, and K. C. Koenen, "Trauma and PTSD in the WHO World Mental Health Surveys," *European Journal of Psychotraumatology* 8, suppl. 5 (2017): 1353383.

5. World Health Organization, "Violence Against Women," https://www.who.int /health-topics/violence-against-women#tab=tab_1.

6. R. Wamser-Nanney and K. E. Cherry, "Children's Trauma-Related Symptoms Following Complex Trauma Exposure: Evidence of Gender Differences," *Child Abuse and Neglect* 77 (2018): 188–97.

7. L. Dell'Osso, C. Carmassi, G. Massimetti, P. Stratta, I. Riccardi, C. Capanna, K. K. Akiskal, H. S. Akiskal, and A. Rossi, "Age, Gender and Epicenter Proximity Effects on Post-traumatic Stress Symptoms in L'Aquila 2009 Earthquake Survivors," *Journal of Affective Disorders* 146, no. 2 (2013): 174–80; C. Carmassi, H. S. Akiskal, D. Bessonov, G. Massimetti, E. Calderani, P. Stratta, A. Rossi, and L. Dell'Osso, "Gender Differences in DSM-5 versus DSM-IV-TR PTSD Prevalence and Criteria Comparison Among 512 Survivors to the L'Aquila Earthquake," *Journal of Affective Disorders* 160 (2014): 55–61.

8. D. M. Christiansen and M. Hansen, "Accounting for Sex Differences in PTSD: A Multi-variable Mediation Model," *European Journal of Psychotraumatology* 6 (2015): 26068.

9. R. C. Kessler, "Epidemiology of Women and Depression," *Journal of Affective Disorders* 74 (2003): 5–13; R. McGee, M. Feehan, S. Williams, and J. Anderson, "DSM-III Disorders from Age 11 to Age 15 Years," *Journal of the American Academy of Child and Adolescent Psychiatry* 31 (1992): 51–59; J. C. Anderson, S. Williams, R. McGee, and P. A. Silva, "DSM-III Disorders in Preadolescent Children: Prevalence in a Large Sample from the General Population," *Archives of General Psychiatry* 44 (1987): 69–77.

10. C. S. Woolley, E. Gould, M. Frankfurt, and B. S. McEwen, "Naturally Occurring Fluctuation in Dendritic Spine Density on Adult Hippocampal Pyramidal Neurons," *Journal of Neuroscience* 10, no. 12 (1990): 4035–39; C. S. Woolley and B. S. McEwen, "Estradiol Mediates Fluctuation in Hippocampal Synapse Density During the Estrous Cycle in the Adult Rat," *Journal of Neuroscience* 12, no. 7 (1992): 2549–54.

11. C. M. Taylor, L. Pritschet, R. K. Olsen, E. Layher, T. Santander, S. T. Grafton, and E. G. Jacobs, "Progesterone Shapes Medial Temporal Lobe Volume Across the Human Menstrual Cycle," *Neuroimage* 2020 (2020): 117125.

12. Robert Sapolsky, personal communication with the author, 2021.

13. Sophia Loren Quotes (2021) BrainyQuote.com, BrainyMedia Inc.

14. B. Leuner and T. J. Shors, "Learning During Motherhood: A Resistance to Stress," *Hormones and Behavior* 50, no. 1 (2006): 38–51.

15. L. Y. Maeng and T. J. Shors, "Once a Mother, Always a Mother: Maternal Experience Protects Females from the Negative Effects of Stress on Learning," *Behavioral Neuroscience* 126, no. 1 (2012): 137–41.

16. J. Milligan-Saville and B. Graham, "Mothers Do It Differently: Reproductive Experience Alters Fear Extinction in Female Rats and Women," *Translational Psychiatry* 6 (2016): e928.

17. E. Hoekzema, E. Barba-Müller, C. Pozzobon, M. Picado, F. Lucco, D. García-García, J. C. Soliva, A. Tobeña, M. Desco, E. A. Crone, A. Ballesteros, S. Carmona, and O. Vilarroya, "Pregnancy Leads to Long-Lasting Changes in Human Brain Structure," *Nature Neuroscience* 20, no. 2 (2017): 287–96.

18. T. J. Shors, E. M. Millon, H. Y. M. Chang, R. L. Olson, and B. L. Alderman, "Do Sex Differences in Rumination Explain Sex Differences in Depression?," *Journal of Neuroscience Research* 95 (2017): 711–18; S. Nolen-Hoeksema and B. Jackson, "Mediators of the Gender Difference in Rumination," *Psychology of Women Quarterly* 25, no. 1 (2001): 37–47.

19. T. J. Shors and E. M. Millon, "Sexual Trauma and the Female Brain," *Frontiers in Neuroendocrinology* 41 (2016): 87–98; E. M. Millon, H. Y. M. Chang, and T. J. Shors, "Stressful Life Memories Relate to Ruminative Thoughts in Women with Sexual Violence History, Irrespective of PTSD," *Frontiers in Psychiatry* 9 (2018): 311.

20. S. Nolen-Hoeksema, "Emotion Regulation and Psychopathology: The Role of Gender," *Annual Review of Clinical Psychology* 8 (2012): 161–87.

21. Shors et al., "Do Sex Differences in Rumination."

22. S. Nolen-Hoeksema and B. Jackson, "Mediators of the Gender Difference"; Nolen-Hoeksema, "Emotion Regulation."

23. General references on topic: J. R. Rainville, T. Lipuma, and G. E. Hodes, "Translating the Transcriptome: Sex Differences in the Mechanisms of Depression and Stress, Revisited," *Biological Psychiatry* (2021). S0006-3223(21)00107-4; Aditi Bhargava, A. P. Arnold, D. A. Bangasser, K. M. Denton, A. Gupta, L. M. H. Krause, E. A. Mayer, M. McCarthy, W. L. Miller, A. Raznahan, and R. Verma, "Considering Sex as a Biological Variable in Basic and Clinical Studies: An Endocrine Society Scientific Statement," *Endocrine Reviews* 42, no. 3 (2021): 219–58; Claire Ainsworth, "Sex Redefined," *Nature* 518 (2015) 288–91; N. Kokras, G. E. Hodes, D. A. Bangasser, and C. Dalla, "Sex Differences in the Hypothalamic-Pituitary-Adrenal Axis: An Obstacle

to Antidepressant Drug Development?," *British Journal of Pharmacology* 176, no. 21 (2019): 4090–106, E. Jazin and L. Cahill, "Sex Differences in Molecular Neuroscience: From Fruit Flies to Humans," *Nature Reviews in Neuroscience* 11 (2010): 9–17.

7. EVERYDAY NEURONS FOR EVERYDAY LIFE

1. M. A. Boldrini, A. N. Santiago, R. Hen, A. J. Dwork, G. B. Rosoklija, H. Tamir, V. Arango, and J. J. Mann, "Hippocampal Granule Neuron Number and Dentate Gyrus Volume in Antidepressant-Treated and Untreated Major Depression," *Neuropsychopharmacology* 38, no. 6 (2013): 1068–77; M. Boldrini, H. Galfalvy, A. J. Dwork, G. B. Rosoklija, I. Trencevska-Ivanovska, G. Pavlovski, R. Hen, V. Arango, and J. J. Mann, "Resilience Is Associated with Larger Dentate Gyrus, While Suicide Decedents with Major Depressive Disorder Have Fewer Granule Neurons," *Biological Psychiatry* 85, no. 10 (2019): 850–62.

2. R. S. Sloviter, G. Valiquette, G. M. Abrams, E. C. Ronk, A. L. Sollas, L. A. Paul, and S. Neubort, "Selective Loss of Hippocampal Granule Cells in the Mature Rat Brain After Adrenalectomy," *Science* 243, no. 4890 (1989): 535–38.

3. J. Altman and G. D. Das, "Autoradiographic and Histological Evidence of Postnatal Hippocampal Neurogenesis in Rats," *Journal of Comparative Neurology* 124, no. 3 (1965): 319–35; M. S. Kaplan and J. W. Hinds, "Neurogenesis in the Adult Rat: Electron Microscopic Analysis of Light Radioautographs," *Science* 197, no. 4308 (1977): 1092–94.

4. H. A. Cameron, C. S. Woolley, B. S. McEwen, and E. Gould, "Differentiation of Newly Born Neurons and Glia in the Dentate Gyrus of the Adult Rat," *Neuroscience* 56, no. 2 (1993): 337–44.

5. T. J. Shors, G. Miesegaes, A. Beylin, M. Zhao, T. Rydel, and E. Gould, "Neurogenesis in the Adult Is Involved in the Formation of Trace Memories," *Nature* 410, no. 6826 (2001): 372–76; T. J. Shors, D. A. Townsend, M. Zhao, Y. Kozorovitskiy, and E. Gould, "Neurogenesis May Relate to Some but Not All Types of Hippocampal-Dependent Learning," *Hippocampus* 12 (2002): 578–84.

6. E. Gould, A. Beylin, P. Tanapat, A. Reeves, and T. J. Shors, "Learning Enhances Adult Neurogenesis in the Hippocampal Formation," *Nature Neuroscience* 2, no. 3 (1999): 260–65.

7. B. Leuner, S. Mendolia-Loffredo, Y. Kozorovitskiy, D. Samburg, E. Gould, and T. J. Shors, "Learning Enhances the Survival of New Neurons Beyond the Time When the Hippocampus Is Required for Memory," *Journal of Neuroscience* 24 (2004) 7477–81; C. Dalla, D. A. Bangasser, C. Edgecomb, and T. J. Shors, "Neurogenesis and Learning: Acquisition and Asymptotic Performance Predict

How Many New Cells Survive in the Hippocampus," *Neurobiology of Learning and Memory* 88 (2007): 143–48; D. M. Curlik II and T. J. Shors, "Learning Increases the Survival of Newborn Neurons Provided That Learning Is Difficult to Achieve and Successful," *Journal of Cognitive Neuroscience* 23, no. 9 (2011): 2159–70.

8. B. Leuner et al., "Learning Enhances the Survival."

9. Y. Kozorovitskiy and E. Gould, "Dominance Hierarchy Influences Adult Neurogenesis in the Dentate Gyrus," *Journal of Neuroscience* 24, no. 30 (2004): 6755–59.

10. Louis D. Matzel, personal communication with the author, circa 1995.

11. D. M. Curlik and T. J. Shors, "Learning Increases the Survival."

12. H. van Praag, G. Kempermann, and F. H. Gage, "Running Increases Cell Proliferation and Neurogenesis in the Adult Mouse Dentate Gyrus," *Nature Neuroscience* 2, no. 3 (1999): 266–70.

13. A. C. Pereira, D. E. Huddleston, A. M. Brickman, A. A. Sosunov, R. Hen, G. M. McKhann, R. Sloan, F. H. Gage, T. R. Brown, and S. A. Small, "An In Vivo Correlate of Exercise-Induced Neurogenesis in the Adult Dentate Gyrus," *Proceedings of the National Academy of Sciences of the United States of America* 104, no. 13 (2007): 5638–43.

14. G. Kempermann, F. H. Gage, L. Aigner, H. Song, M. A. Curtis, S. Thuret, H. G. Kuhn, S. Jessberger, P. W. Frankland, H. A. Cameron, E. Gould, R. Hen, D. N. Abrous, N. Toni, A. F. Schinder, X. Zhao, P. J. Lucassen, and J. Frisén, "Human Adult Neurogenesis: Evidence and Remaining Questions," *Cell Stem Cell* 23, no. 1 (2018): 25–30; H. Lee and S. Thuret, "Adult Human Hippocampal Neurogenesis: Controversy and Evidence," *Trends in Molecular Medicine* 24, no. 6 (2018): 521–22.

15. P. S. Eriksson, E. Perfilieva, T. Björk-Eriksson, A. M. Alborn, C. Nordborg, D. A. Peterson, and F. H. Gage, "Neurogenesis in the Adult Human Hippocampus," *Nature Medicine* 4, no. 11 (1998): 1313–17.

16. E. P. Moreno-Jiménez, M. Flor-García, J. Terreros-Roncal, A. Rábano, F. Cafini, N. Pallas-Bazarra, J. Ávila, and M. Llorens-Martín, "Adult Hippocampal Neurogenesis Is Abundant in Neurologically Healthy Subjects and Drops Sharply in Patients with Alzheimer's Disease," *Nature Medicine* 25, no. 4 (2019): 554–60.

17. K. Fabel, S. A. Wolf, D. Ehninger, H. Babu, P. Leal-Galicia, and G. Kempermann, "Additive Effects of Physical Exercise and Environmental Enrichment on Adult Hippocampal Neurogenesis in Mice," *Frontiers in Neuroscience* 3 (2009): 50.

18. D. M. Curlik II, L. Y. Maeng, P. R. Agarwal, and T. J. Shors, "Physical Skill Training Increases the Number of Surviving New Cells in the Adult Hippocampus," *PLoS One* 8, no. 2 (2013): e55850; G. DiFeo and T. J. Shors, "Mental and Physical Skill Training Increases Neurogenesis via Cell Survival in the Adolescent Hippocampus," *Brain Research* 1654, pt. B (2017): 95–101.

19. R. A. Bjork, J. Dunlosky, and N. Kornell, "Self-Regulated Learning: Beliefs, Techniques, and Illusions," *Annual Review of Psychology* 64 (2013): 417–44.

20. H. M. Sisti, A. L. Glass, and T. J. Shors, "Neurogenesis and the Spacing Effect: Learning Over Time Enhances Memory and the Survival of New Neurons," *Learning and Memory* 14, no. 5 (2007): 368–75; M. S. Nokia, H. M. Sisti, M. R. Choksi, and T. J. Shors, "Learning to Learn: Theta Oscillations Predict New Learning, Which Enhances Related Learning and Neurogenesis," *PLoS One* 7, no. 2 (2012): 31375.

21. K. A. Ericsson, K. Nandagopal, and R. W. Roring, "Toward a Science of Exceptional Achievement: Attaining Superior Performance Through Deliberate Practice," *Annals of the New York Academy of Sciences* 1172 (2009): 199–217.

22. M. Boldrini et al., "Hippocampal Granule,"; M. Boldrini et al., "Resilience Is Associated."

8. THERAPIES FOR STRESS AND TRAUMA

1. L. E. Watkins, K. R. Sprang, and B. O. Rothbaum, "Treating PTSD: A Review of Evidence-Based Psychotherapy Interventions," *Frontiers in Behavioral Neuroscience* 12 (2018): 258.

2. E. B. Foa and M. J. Kozak, "Emotional Processing of Fear: Exposure to Corrective Information," *Psychological Bulletin* 99, no. 1 (1986): 20–35; Barbara Rothbaum, Edna Foa, and Elizabeth Hembree, *Reclaiming Your Life from a Traumatic Experience: A Prolonged Exposure Treatment Program* (Oxford: Oxford University Press, 2007); C. P. McLean and E. B. Foa, "Prolonged Exposure Therapy for Post-traumatic Stress Disorder: A Review of Evidence and Dissemination," *Expert Review of Neurotherapeutics* 11, no. 8 (2011): 1151–63; S. Markowitz, and M. Fanselow, "Exposure Therapy for Post-traumatic Stress Disorder: Factors of Limited Success and Possible Alternative Treatment," *Brain Sciences* 10, no. 3 (2020): 167.

3. Patricia A. Resick, Candice M. Monson, and Kathleen M. Chard, *Cognitive Processing for PTSD: A Comprehensive Manual* (New York: Guilford Press, 2016).

4. P. A. Resick, P. Nishith, T. L. Weaver, M. C. Astin, and C. A. Feuer, "A Comparison of Cognitive-Processing Therapy with Prolonged Exposure and a Waiting Condition for the Treatment of Chronic Posttraumatic Stress Disorder in Female Rape Victims," *Journal of Consulting and Clinical Psychology* 70, no. 4 (2002): 867–79.

5. Jon Kabat-Zinn, *Full Catastrophe Living: Using the Wisdom of Your Body and Mind to Face Stress, Pain, and Illness* (New York: Bantam Books, 1990); J. J. Miller, K. Fletcher, and J. Kabat-Zinn, "Three-Year Follow-Up and Clinical Implications of a Mindfulness Meditation-Based Stress Reduction Intervention in the Treatment of Anxiety Disorders," *General Hospital Psychiatry* 17, no. 3 (1995): 192–200;

J. Taylor, L. McLean, A. Korner, E. Stratton, and N. Glozier, "Mindfulness and Yoga for Psychological Trauma: Systematic Review and Meta-analysis," *Journal of Trauma and Dissociation* 21 (2020): 536–73.

6. Daniel Goleman and Richard J. Davidson, *Altered Traits: Science Reveals How Meditation Changes Your Mind, Brain, and Body* (New York: Avery, 2018); Zindel V. Segal, J. Mark G. Williams, and John D. Teasdale, *Mindfulness-Based Cognitive Therapy for Depression* (New York: The Guildford Press, 2013).

7. L. Santarelli, M. Saxe, C. Gross, A. Surget, F. Battaglia, S. Dulawa, N. Weisstaub, J. Lee, R. Duman, O. Arancio, C. Belzung, and R. Hen, "Requirement of Hippocampal Neurogenesis for the Behavioral Effects of Antidepressants," *Science* 301, no. 5634 (2003): 805–09; J. E. Malberg, A. J. Eisch, E. J. Nestler, and R. S. Duman, "Chronic Antidepressant Treatment Increases Neurogenesis in Adult Rat Hippocampus," *Journal of Neuroscience* 20, no. 24 (2000): 9104–10; G. E. Hodes, L. Yang, J. Van Kooy, J. Santollo, and T. J. Shors, "Prozac During Puberty: Distinctive Effects on Neurogenesis as a Function of Age and Sex," *Neuroscience* 163, no. 2 (2009): 609–17.

8. M. E. Bowers and K. J. Ressler, "An Overview of Translationally Informed Treatments for Posttraumatic Stress Disorder: Animal Models of Pavlovian Fear Conditioning to Human Clinical Trials," *Biological Psychiatry* 78, no. 5 (2015): E15–27.

9. C. G. Fairburn and V. Patel, "The Impact of Digital Technology on Psychological Treatments and Their Dissemination," *Behaviour Research and Therapy* 88 (2017): 19–25.

9. *MAP TRAIN MY BRAIN:* A "MENTAL AND PHYSICAL" TRAINING PROGRAM

1. T. J. Shors, "Saving New Brain Cells," *Scientific American* (March 2009): 46–52.

2. David Chadwick, *Crooked Cucumber: The Life and Zen Teaching of Shunryu Suzuki* (New York: Broadway Books, 1999), 301.

3. D. Attwell and S. B. Laughlin, "An Energy Budget for Signaling in the Grey Matter of the Brain," *Journal of Cerebral Blood Flow and Metabolism* 21 (2001): 1133–45.

4. K. Van der Borght, D. E. Kóbor-Nyakas, K. Klauke, B. J. L. Eggen, C. Nyakas, E. A. Van der Zee, and P. Meerlo, "Physical Exercise Leads to Rapid Adaptations in Hippocampal Vasculature: Temporal Dynamics and Relationship to Cell Proliferation and Neurogenesis," *Hippocampus* 19 (2009): 928–36.

5. G. R. Wylie, J. J. Foxe, and T. L. Taylor, "Forgetting as an Active Process: An fMRI Investigation of Item-Method-Directed Forgetting," *Cerebral Cortex* 18, no. 3 (2008): 670–82.

6. S. R. Chekroud, R. Gueorguieva, A. B. Zheutlin, M. Paulus, H. M. Krumholz, J. H. Krystal, and A. M. Chekroud, "Association Between Physical Exercise and Mental Health in 1.2 Million Individuals in the USA Between 2011 and 2015: A Cross-Sectional Study," *Lancet Psychiatry* 5, no. 9 (2018): 739–46.

7. E. Zotcheva, C. W. S. Pintzka, Ø. Salvesen, G. Selbæk, A. K. Håberg, and L. Ernstsen, "Associations of Changes in Cardiorespiratory Fitness and Symptoms of Anxiety and Depression with Brain Volumes: The HUNT Study," *Frontiers in Behavioral Neuroscience* 13 (2019): 53.

8. L. D. Baker, L. L. Frank, K. Foster-Schubert, P. S. Green, C. W. Wilkinson, A. McTiernan, S. R. Plymate, M. A. Fishel, G. S. Watson, B. A. Cholerton, G. E. Duncan, P. D. Mehta, and S. Craft, "Effects of Aerobic Exercise on Mild Cognitive Impairment: A Controlled Trial," *Archives of Neurology* 67, no. 1 (2010): 71–79.

9. J. Najar, S. Östling, P. Gudmundsson, V. Sundh, L. Johansson, S. Kern, X. Guo, T. Hällström, and I. Skoog, "Cognitive and Physical Activity and Dementia: A 44-Year Longitudinal Population Study of Women," *Neurology* 92, no. 12 (2019): e1322–e1330.

10. WHY WE SHOULD TRAIN OUR BRAINS

1. D. S. Hasin, A. L. Sarvet, J. L. Meyers, T. D. Saha, W. J. Ruan, M. Stohl, and B. F. Grant, "Epidemiology of Adult *DSM-5* Major Depressive Disorder and Its Specifiers in the United States," *JAMA Psychiatry* 75, no. 4 (2018): 336–46.

2. B. L. Alderman, R. L. Olson, C. J. Brush, and T. J. Shors, "MAP Training: Combining Meditation and Aerobic Exercise Reduces Depression and Rumination While Enhancing Synchronized Brain Activity," *Translational Psychiatry* 6, no. 2 (2016): e726.

3. Ibid.

4. T. J. Shors, R. L. Olson, M. E. Bates, E. A. Selby, and B. L. Alderman, "Mental and Physical (MAP) Training: A Neurogenesis-Inspired Intervention That Enhances Brain Health in Humans," *Neurobiology of Learning and Memory* 115 (2014): 3–9.

5. T. J. Shors, H. Y. M. Chang, and E. M. Millon, "*MAP Training My Brain*: Meditation Plus Aerobic Exercise Lessens Trauma of Sexual Violence More Than Either Activity Alone," *Frontiers in Neuroscience* 12 (2018): 211.

6. Shors et al., "Mental and Physical (MAP) Training"; E. M. Millon and T. J. Shors, "Taking Neurogenesis out of the Lab and into the World with *MAP Train My Brain*," *Behavioural Brain Research* 376 (2019): 112154.

7. P. Lavadera, E. M. Millon, and T. J. Shors, "*MAP Train My Brain*: Meditation Combined with Aerobic Exercise Reduces Stress and Rumination While

Enhancing Quality of Life in Medical Students," *Journal of Alternative and Complementary Medicine* 26, no. 5 (2020): 418–23.

8. S. J. H. van Rooij, M. Ravi, T. D. Ely, V. Michopoulos, S. J. Winters, J. Shin, M. F. Marin, M. R. Milad, B. O. Rothbaum, K. J. Ressler, T. Jovanovic, and J. S. Stevens, "Hippocampal Activation During Contextual Fear Inhibition Related to Resilience in the Early Aftermath of Trauma," *Behavioural Brain Research* 408 (2021): 113282; M. I. Nash, C. B. Hodges, N. M. Muncy, and C. B. Kirwan, "Pattern Separation Beyond the Hippocampus: A High-Resolution Whole-Brain Investigation of Mnemonic Discrimination in Healthy Adults," *Hippocampus* 31, no. 4 (2020): 408–21.

9. General references on topic: D. Marr, "Simple Memory: A Theory for Archicortex," *Philosophical Transactions of the Royal Society B* 262, no. 841 (1971): 23–81; S. Becker, "A Computational Principle for Hippocampal Learning and Neurogenesis," *Hippocampus* 15, no. 6 (2005): 722–38; A. Bakker, C. B. Kirwan, M. Miller, and C. E. Stark, "Pattern Separation in the Human Hippocampal CA3 and Dentate Gyrus," *Science* 319, no. 5870 (2008): 1640–42; C. D. Clelland, M. Choi, C. Romberg, G. D. Clemenson Jr., A. Fragniere, P. Tyers, S. Jessberger, L. M. Saksida, R. A. Barker, F. H. Gage, and T. J. Bussey, "A Functional Role for Adult Hippocampal Neurogenesis in Spatial Pattern Separation," *Science* 325 (2009): 210–13; A. Sahay, N. Scobie, A. S. Hill, C. M. O'Carroll, M. A. Kheirbek, N. S. Burghardt, A. A. Fenton, D. Dranovsky, and R. Hen, "Increasing Adult Hippocampal Neurogenesis Is Sufficient to Improve Pattern Separation," *Nature* 472, no. 7344 (2011): 466–70; M. A. Yassa and C. E. Stark, "Pattern Separation in the Hippocampus," *Trends in Neurosciences* 34, no. 10 (2011): 515–25.

10. E. M. Millon, P. Lehrer, and T. J. Shors, "Mental and Physical (MAP) Training with Meditation and Aerobic Exercise Enhances Mental Health Outcomes and Pattern Separation in Women Living with HIV," (2021, under review for publication).

11. LIVING WITH TRAUMAS: PAST, PRESENT, AND FUTURE

1. William James, *Principles of Psychology* (New York: Henry Holt, 1918), Chapter 4.

2. H. Kim, H. R. Smolker, L. L. Smith, M. T. Banich, and J. A. Lewis-Peacock, "Changes to Information in Working Memory Depend on Distinct Removal Operations," *Nature Communications* 11, no. 1 (2020): 6239; M. T. Banich, K. L. M. Seghete, B. E. Depue, and G. C. Burgess, "Multiple Modes of Clearing One's Mind of Current Thoughts: Overlapping and Distinct Neural Systems," *Neuropsychologia* 69 (2015): 105–17.

3. S. R. Chamberlain, J. E. Grant, W. Trender, P. Hellyer, and A. Hampshire, "Post-traumatic Stress Disorder Symptoms in COVID-19 Survivors: Online Population Survey," *British Journal of Psychiatry* 7, no. 2 (2021): e47; M. Taquet, J. R. Geddes, M. Husain, S. Luciano, and P. J. Harrison, "6-Month Neurological and Psychiatric Outcomes in 236,379 Survivors of COVID-19: A Retrospective Cohort Study Using Electronic Health Records," *Lancet Psychiatry* 8 no. 5 (2021): 416–27.

4. D. Demmin, S. M. Silverstein, and T. J. Shors, "Mental and Physical (MAP) Training with Meditation and Aerobic Exercise During the COVID-19 Pandemic Reduces Stress and Improves Well-Being in Teachers," (2021, under review for publication).

INDEX

ABOUT THE AUTHOR

Tracey J. Shors, Ph.D., is a distinguished professor in behavioral and systems neuroscience and a member of the Center for Collaborative Neuroscience at Rutgers University. She is also vice chair and director of graduate studies in the department of psychology. Dr. Shors has published more than 140 scientific articles, including reports in *Nature, Nature Neuroscience, Proceedings of the National Academy of Sciences of the United States of America,* and *Science,* and her work has been featured in *Scientific American, The New York Times, The Washington Post,* and on NPR and CNN. She has been at Rutgers University for more than twenty years.